AMERICAN TROUT-STREAM INSECTS

EXPERT DRY FLY-CASTING ON THE STREAM
(Pencil Portrait of Mr. George La Branche)

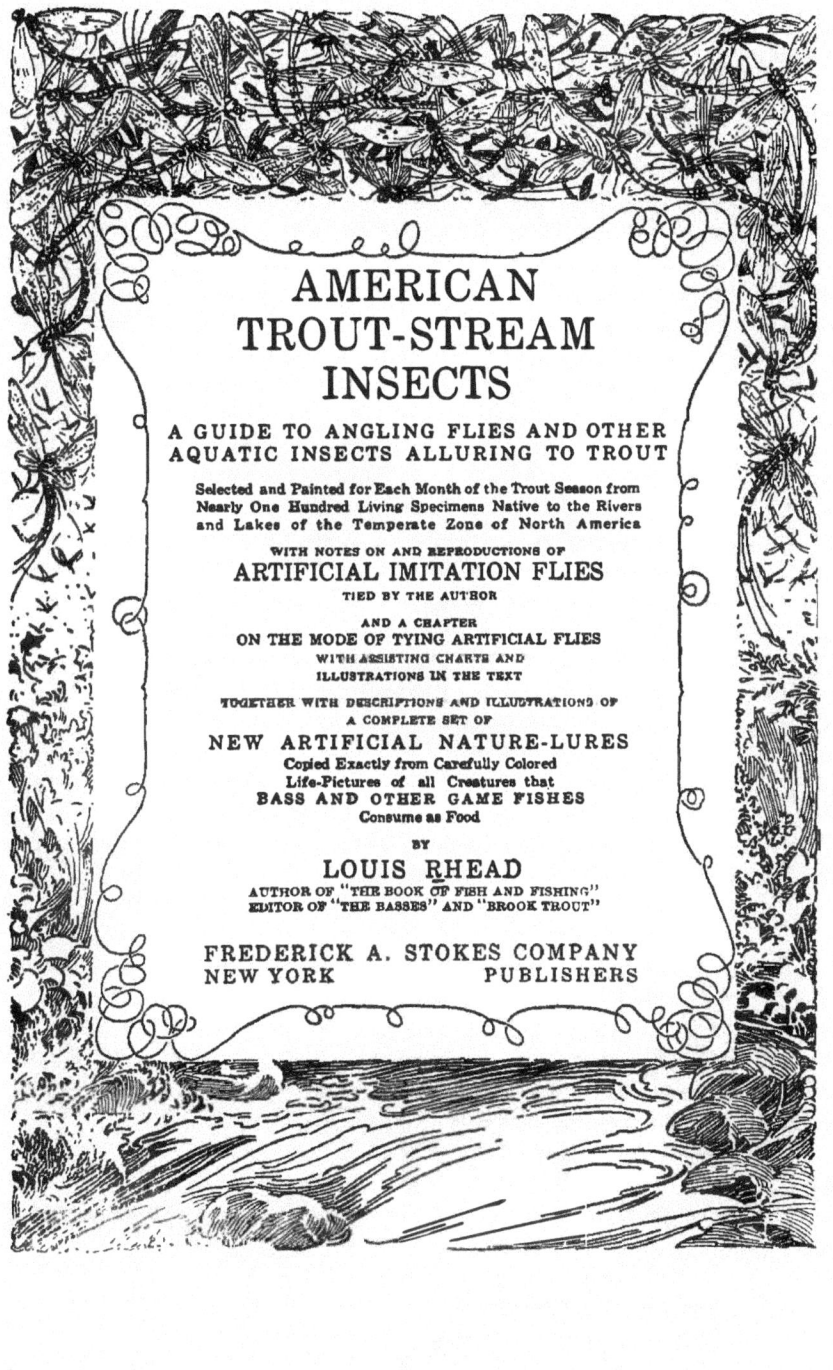

AMERICAN TROUT-STREAM INSECTS

A GUIDE TO ANGLING FLIES AND OTHER
AQUATIC INSECTS ALLURING TO TROUT

Selected and Painted for Each Month of the Trout Season from
Nearly One Hundred Living Specimens Native to the Rivers
and Lakes of the Temperate Zone of North America

WITH NOTES ON AND REPRODUCTIONS OF
ARTIFICIAL IMITATION FLIES
TIED BY THE AUTHOR

AND A CHAPTER
ON THE MODE OF TYING ARTIFICIAL FLIES
WITH ASSISTING CHARTS AND
ILLUSTRATIONS IN THE TEXT

TOGETHER WITH DESCRIPTIONS AND ILLUSTRATIONS OF
A COMPLETE SET OF
NEW ARTIFICIAL NATURE-LURES
Copied Exactly from Carefully Colored
Life-Pictures of all Creatures that
BASS AND OTHER GAME FISHES
Consume as Food

BY
LOUIS RHEAD
AUTHOR OF "THE BOOK OF FISH AND FISHING"
EDITOR OF "THE BASSES" AND "BROOK TROUT"

FREDERICK A. STOKES COMPANY
NEW YORK   PUBLISHERS

ISBN 978-1-4341-0548-6

Published by Waking Lion Press, an imprint of the Editorium

Waking Lion Press™ and Editorium™ are trademarks of:

The Editorium, LLC
West Jordan, UT 84081-6132
www.editorium.com

The views expressed in this book are the responsibility of the author and do not necessarily represent the position of Waking Lion Press. The reader alone is responsible for the use of any ideas or information provided by this book.

TO MY HONORED
AND ESTEEMED FRIEND OF MANY YEARS

TARLETON H. BEAN

Fish Culturist of the State of New York
President of the American Fisheries Society
Founder and First Director of the New York Aquarium
Chief of the Department of Fish and Game at Many Universal Expositions
Author of Innumerable and Important Piscicultural Books and Government Documents
Decorated in the Legion of Honor and Mérite Agricole of France the Red Eagle of Germany, and the Rising Sun of Japan for his Services to Fish Culture and the Fisheries,
etc., etc.

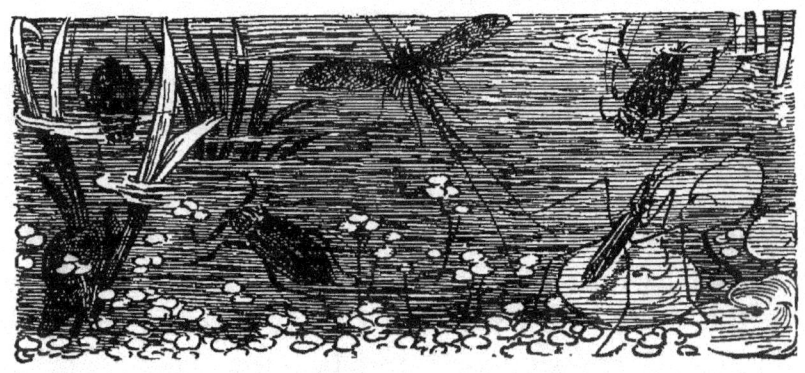

PREFACE

THE object of this book is to furnish anglers, amateurs, students of entomology, and others interested in aquatic insects, with a colored selection of the most abundant and well-known trout insects that appear, month by month, on the rivers and lakes of the temperate regions of North America.

Inquiries from various State entomologists failed to locate a single volume or treatise on trout-stream insects. Likewise, diligent search in libraries and large bookstores proved futile. This shows the present volume to be the first and only work on this most necessary adjunct to the angler's craft. It is curious that while our Government documents relating to pisciculture are equal to, if not in advance of, those of European countries, there, particularly in England and France, research in the field of entomology has kept pace with other branches of science.

After looking back over a period of seven trout

PREFACE

seasons, in which the insect studies of this book have been in preparation, the author well understands why the subject has not been treated heretofore in America. No artist could do it properly unless he were an angler of wide experience, with an infinite love of his subject, and willing to devote ample time to the work. He must be present on the streams nearly all the summer; and he must be prepared to forego much pleasure in the pursuit of his favorite sport—it so often happens trout are rising to the very flies which must needs be captured. No angler could do it unless he were an artist; and no artist unless he were an angler; it must be a combination of both.

The difficulty in catching, uninjured, these most fragile insect specimens, in keeping them alive in a wire cage long enough to paint them in colors true to the *living* fly (when dead, their beautiful color instantly fades), is an undertaking my brother anglers would scarcely believe. Most of the insects must, of necessity, be captured at evening—very often miles away from home. It is imperative that the captives be kept over night in the open air; and in the morning many of the most delicate are either dead or so greatly injured as to be useless. Then the hunt for them must be gone all over again. Nearly every specimen has to be painted by the aid of a magnifying-glass; and the most fragile are the most restless. To stick a pin

PREFACE

in them is sure death, the mangling of the body, and the fading of the color.

Aquatic insects—at least, those most alluring to trout—have a tendency to keep themselves hovering over deep or rushing water, while the maddened would-be capturer stands waiting to get them within reach of his net, or else, teasingly tempted, he flounders about in the water, wet to the skin, only to miss the object of his desire.

Worse, far worse, are these elusive insects to get within the net than the wily trout.

Louis Rhead.

Brooklyn, N. Y.

CONTENTS

PART I

AMERICAN TROUT-STREAM INSECTS

CHAPTER		PAGE
	Preface	vii
	Introductory Note	xv
I	Why It Is Best to Copy Nature	1
II	System and Classification	5
III	The Artificial Fly of Commerce	13
IV	Trout Flies in April—When Insects First Appear	18
V	Trout Insects for May	35
VI	The Best Trout Insects for June	52
VII	Typical Insects of July	66
VIII	Some Trout Insects for August	77
IX	Six Best Flies for Each Month	87
X	Concerning My Artificial Imitations	97
XI	New Names for Flies	101
XII	The Making of an Artificial Fly	104
XIII	A Test of the New Flies	124

CONTENTS

PART II

NEW ARTIFICIAL NATURE LURES

CHAPTER		PAGE
	INTRODUCTORY NOTE: CONCERNING THE RAPIDLY DIMINISHING NATURAL FOOD OF FRESH-WATER GAME FISHES	129
XIV	NEW LURES THAT ARE TRUE TO LIFE . . .	133
XV	SHINY DEVILS: GOLD- AND SILVER-BODIED FANCY MINNOWS FOR SALMON, BIG TROUT, AND BASS .	139
XVI	NATURE LURES FOR SUMMER FISHING: IMITATIONS OF MINNOWS, GRASSHOPPERS, DRAGONFLIES AND CATERPILLARS	144
XVII	ARTIFICIAL FROGS THAT WIGGLE THEIR LEGS AND FLOAT	151
XVIII	THE THREE BEST NATURE LURES FOR BASS . .	158
XIX	SILVER SHINER AND GOLDEN CHUB: NEW MINNOWS FOR SURFACE OR DEEP-WATER TROLLING AFTER BIG LAKE TROUT, TOGUE, MASCALONGE, OR SALMON	164
XX	THE RIGHT IMPLEMENTS AND METHODS: THEIR IMPORTANCE FOR SURE SUCCESS	173

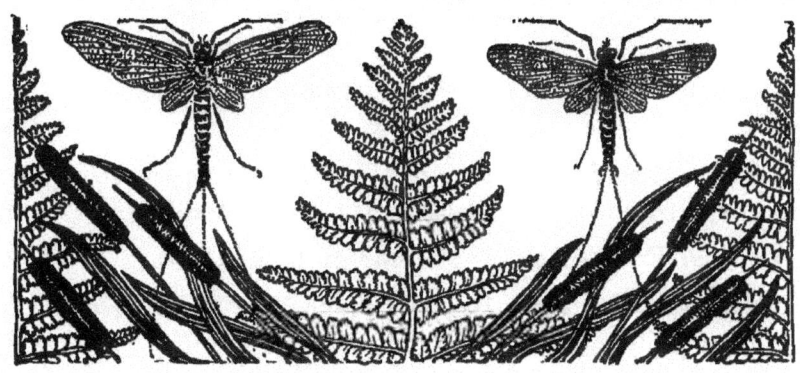

ILLUSTRATIONS

Expert Dry Fly-Casting on the Stream (Pencil Portrait of Mr. George La Branche). *In colors* . . . . *Frontispiece*

FACING PAGE

Specimens of Seven Different Orders of Insects . . . 6

A Selection of the Best Trout Insects for the Month of April. Painted from Life by the Author. *In colors* . 20

A Selection of the Best Trout Insects for the Month of May. Painted from Life by the Author. *In colors* . 36

A Selection of the Best Trout Insects for the Month of June. Painted from Life by the Author. *In colors* . 54

A Selection of the Best Trout Insects for the Month of July. Painted from Life by the Author. *In colors* . 68

A Selection of the Best Trout Insects for the Month of August and Corresponding Artificial Flies Tied by the Author 80

Selected Nature Flies Tied in Accordance with the Author's Patterns and Sold by His Agents 90

Tools for, and Methods of Making a Fly 106

Feather Minnows for Bass, Pike and Trout 134

ILLUSTRATIONS

	FACING PAGE
Gold- and Silver-Bodied Fancy Minnows for Salmon, Big Trout and Bass	140
Nature Lures for Summer Fishing: Imitations of Minnows, Grasshoppers, Dragon Fly and Caterpillars	144
Artificial Frogs—Lamper Eel and Helgramite	152
Crawfish: Under View, Back View and Side View. Helgramite: Side View	158
Silver Shiner (actual size) for either Surface or Deep Water Trolling after Lake Trout, Mascalonge, Pike and Salmon	166

INTRODUCTORY NOTE

OVER twenty years ago Mary Orvis Marbury, in her well-known and useful book, "Favorite Flies," wrote: "At present fishermen are chiefly indebted to the fly-makers of Great Britain for copies of the insects alluring to game fish . . . and until we have studied more thoroughly our own stream life we do well to abide by many of their conclusions; but there can be no question that in the years to come the differences between the insects of the two countries will be better understood and defined, and that a collection of the water insects interesting to fishermen of America, with directions for accurate imitations, arranged after the manner of Ronalds' 'Fly-fisher's Entomology,' would be of great value."

To-day conditions remain the same. No one, as yet, has seen fit to undertake such a collection of American insects; although in the interval several important books on English insects have been issued—notably, Halford's "Entomology" and

INTRODUCTORY NOTE

Leonard West's "The Natural Trout Fly and Its Imitations."

Now that so many American fly-fishermen desire a knowledge of this subject, it would seem that the time has arrived for a book on American insects; from the fact that anglers in all parts of the country have requested me to undertake this much-needed work.

In this volume I purpose to describe and picture in colors a selected number of the most abundant and most common insects that trout feed upon in a typical American trout stream, and to show, side by side with these, correct artificial imitation flies tied by my own hands, in order that anglers may better understand how to choose their own flies and thus be enabled to lure fish with greater success and pleasure than heretofore.

In a handbook of this kind it would not be possible to include all of the large number of species and the numerous varieties that inhabit the different trout waters of the United States. I shall describe those specimens found on streams in the mountainous regions of New York and Pennsylvania, which are available to the more northern and perhaps southern zones.

At the present time American and English artificial flies are, almost exclusively, what may be called "fancy flies"; that is, tied flies made of colored feathers and wool, without much attempt at imi-

INTRODUCTORY NOTE

tating the natural insects. The question may possibly be asked, "If fancy flies entice trout, is not the object for which they are intended attained?" My reply is, good imitations of the insects that trout take as food are sure to prove more effective and make far more interesting sport.

The great majority of expert anglers at the present time rely on fly-makers of the British Isles for their trout and salmon flies; moreover, we are naïvely assured, "Some are tied especially for American waters." In the various plates of colored flies, the angler will see a much greater difference between the insects of the two countries than is commonly supposed to exist; though in a few instances there are species that are nearly alike—as for instance the green and brown drake. It is most essential that a true copy of our native insects should be used as a guide for American anglers.

European entomologists have divided insects into various orders; each season finds them making new classifications so conflicting as to bewilder the lay mind. For the simple use of plain anglers, who have neither the wish nor the time to enter into the intricacies of entomology, the use of a common name for each insect will, I think, be more acceptable. Among the insects pictured will be found specimens of browns, drakes, duns, spinners, beetles, house-flies and ants. To enter into details concern-

INTRODUCTORY NOTE

ing the general classification of insects would prove wearisome and perplexing and would destroy the purpose intended. This book endeavors, merely, to simplify the branch relating only to insects of interest to anglers.

The insects pictured are carefully selected from the many captured in the past three years, while wading and fishing in the river Beaverkill, situated in the Catskill region of New York. They have been most carefully copied, in every instance, from the *living* insects, which, like fish, rapidly change their lovely color soon after death. They represent what I consider are the most valuable insects to imitate that will most induce trout to take them.

I have tied the various artificials as I think they will best imitate the natural insects, without any reference whatever to the artificial flies made and sold by American and English fly-makers, although I have made a careful study of all the works on trout flies. I have no hesitation in pinning my faith on the angler who uses these new patterns, if used as directed in the accompanying charts, as against the expert who angles with the popular native "fancy" flies or even the imported English dry flies.

Of this I am sure: for every insect a trout takes alive at the surface, a thousand are consumed drowned under water or near the surface; and to

INTRODUCTORY NOTE

one natural insect able to float on the surface, there are hundreds which cannot float.

Because of this fact, I believe it to be the height of folly to fish exclusively with dry flies on the surface. Wet fishing with two or three accurately copied insects is in every way as effective on the average American stream. I do think the dry-fly method is excellent on large pools, and more or less placid water; but the trouble is that trout prefer to lie under a rock where turbulent water flows above, from which in a runway they get insects alive or drowned as they go swiftly by.

Since this appeared in magazine form I have read Mr. La Branche's "Dry Fly in Rapid Waters," and I quite agree that anglers can wade into the lower end of a rapid, and with a very short cast float down a cocked dry fly a few yards toward them with deadly effect. I have seen Mr. La Branche do it, and I have done it frequently myself.

In general fishing the method is merely a matter of preference, and is really of very little importance compared to offering the right artificial that will make trout fancy is its regular food.

PART I
AMERICAN TROUT-STREAM INSECTS

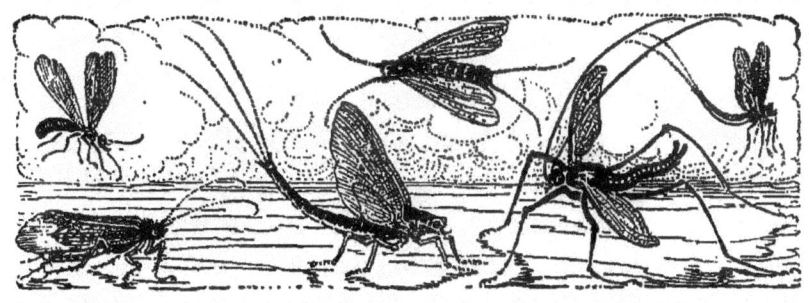

AMERICAN TROUT-STREAM INSECTS

I

WHY IT IS BEST TO COPY NATURE

EVERY thoughtful angler will agree that to fish with an exact artificial imitation of the natural insect is certainly a desirable thing. To do this it will be necessary that some radical changes be made from the fly of commerce now in use.

In the first place, the most important class, called drakes—and a great many specimens of the other classes—have their under bodies quite pale and colorless compared to the upper part of their bodies. From the trout's point of view—that is, looking upward—the artificial imitation with pale or white under body must be a much more acceptable lure than the commercial fly as now tied with upper and under body exactly the same.

Another desirable change, still more important, is to tie all drakes' wings close together instead of outspread—a most unnatural attitude which the insect never assumes, and only introduced in later years by dry fly experts in order to assist the fly in floating. With the aid of oil the fly will float just as well with closed wings as with wings outspread.

One of the most remarkable and peculiar features in all drakes is the way they cock their tails upward from the body. In order to imitate this feature successfully it is necessary to have what is called a "detached" body: viz., with the tail made separate from the hook and slightly curved upward. There are a few English flies tied in this way; but I have not seen any detached bodies on what may be called strictly American flies. The detached-body fly I have found far ahead of the fly with the body tied round the hook: that is, if intended to represent the drake class. For that reason I have tied all my drakes with detached bodies.

Still another change for the better is to tie the numerous class of duns with wings sloping down over the body like the natural insect. There are many hundreds of these aquatic insects which, after alighting on the water's surface, rest with the wings lapping over the covering of the body; which is exactly opposite to the raised wings of commercial flies.

It will therefore be seen that the two principal

WHY IT IS BEST TO COPY NATURE

classes—duns and drakes—have wings in exactly opposite positions. The duns have wings hanging from and below the shoulders; the drakes, wings raised from and erect above the shoulders.

Trout fishermen also would find their sport vastly improved if they used imitations of the large and important class of spinners, which form a considerable part of the trout's food. So far as I know, no attempt has been made to copy spinners. The extra long hackle used for the legs, the long, thin body, and the spread wings must surely act as an effective teaser when lightly dropped and floated to large trout. They might be excellent, also, for river bass.

I had no idea that the spinners were so numerous in numbers and variety. That they have not been imitated before this and used as a regular commercial fly seems strange to me. I see trout gobble them whenever they touch the surface; yes, and even jump at the surface of the water after them while on the wing.

If you fish with an artificial fly like the insect you see the trout takes it gives additional vim to the sport. Your interest is doubled, not only because it is certain that trout will respond to correct imitations, but by the charm of the situation whereby you are as intimately connected with the food taken as with the fish.

By using a fancy fly your interest in what food

the fish prefers is void; you simply fish with a bunch of feathers. Furthermore, it may be necessary to keep changing fancy flies till you get the desired rise, which may not come to any of them.

How often, when a nice fish is netted, is the question asked, "What fly did you use?"

I claim, and have proved many times, that a fairly good imitation of rising insects will induce trout to rise when a fancy fly is ineffective. If an exact copy of the natural insect is offered to the fish—even if that natural insect is not in flight—it is sure to entice and lure a trout more readily than a fancy fly.

That is the reason this book of the commonest and most abundant trout insects is offered for the use of discerning and thoughtful anglers.

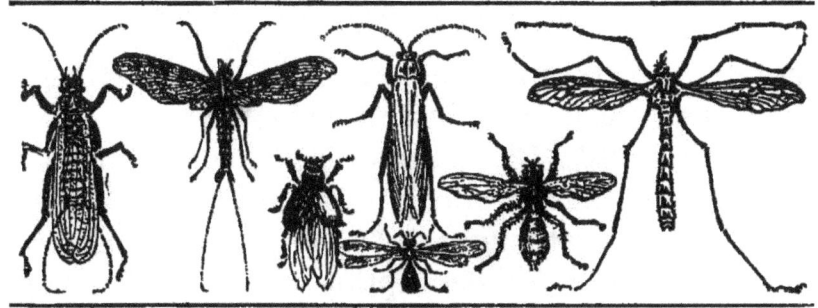

II

SYSTEM AND CLASSIFICATION

AFTER a careful study of the various British books on trout insects and their artificial imitations issued to date, I deem it wise to brush aside the science of entomology, which is of no actual service to our purpose, and to lay before the angler a plain, simple plan whereby he can obtain just enough information to understand easily the general characteristics of the insects he is likely to observe trout feeding upon while wading a trout stream, to the end that he may have with him a fair imitation that will be most successful in luring trout.

For what little classification is necessary I have, in a measure, followed after the plan of Michael Theakston, who was the first British angler to make an effort in that direction. He also made a selected list of the best British flies for each month, and tied his flies true to nature. The wings of his drakes are cocked; those of his browns and duns are flat. You can at once recognize in his artificial imitations the class of insects he imitates.

In Ronalds' "Fly-fisher's Entomology" there is no effort at classification; he simply made a selection of what he thought to be the best flies for each month, giving a full description of them and telling how to imitate them. Ronalds was an exceptionally clever artist, and his representations of both insects and artificials are so perfect that I form from them a better understanding of British aquatic insects than from any others I have seen issued up to the present time. He, like Theakston, tied his own flies with the greatest fidelity to nature.

The plans of these two eminent authorities I consider are the best to follow; for every month of our trout season has a distinctly different weather condition, and, too, there is a decided difference in our trout insects each month.

The accompanying page drawing shows a system of separation according to shape and construction of the seven most important classes of trout insects and their creepers, viz.: browns, drakes, duns, spinners, beetles, house-flies, and ants. Each of these classes contains many distinct species that differ in size, color, and shape. The monthly colored pages show only a selection of the best and most abundant insects seen upon the water.

Take, for instance, the order of *Ephemera,* a

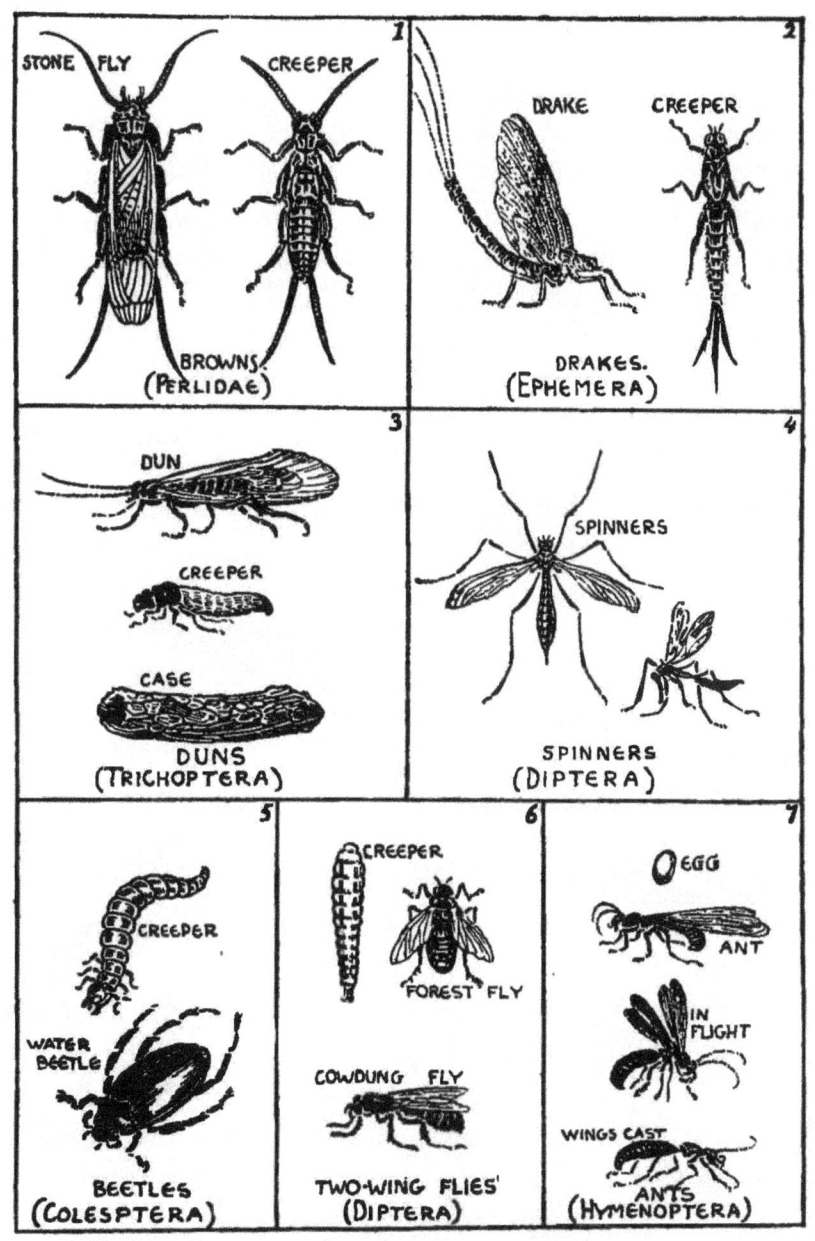

SPECIMENS OF SEVEN DIFFERENT ORDERS OF INSECTS

drake class, of which there are over forty species in the British Isles. This region somewhat corresponds to our temperate regions in climatic condition, yet I am pretty sure that not only have we many more species of that class but they are larger, much more brilliant and varied in color. The same will no doubt be found true with the class of *Perlidæ,* or stone-flies, and other orders.

I have not yet had the opportunity to study the insects north or south of the temperate regions; yet I am of the opinion that both those sections of our American continent will show an entirely different class of insects. A gentleman from North Carolina kindly sent me some trout flies caught in his locality which were entirely new to me; one, in particular, a spinner of gigantic proportions compared with those found here.

The following table of seven orders the angler will be able to use as a guide to identify different species in their class whenever he observes them in their natural state, either as creeper or as mature insect:

First: The browns, or stone-flies, are all bred in the water and hatch out from a creeper. They are most plentiful in the spring, but some are on the water all summer. Their bodies are smooth and fleshy, and they have two pairs of smooth, oblong wings, which, when folded, circle closely over and beyond the body. They have two feelers at

SYSTEM AND CLASSIFICATION

the head, and most of them two wisks at the tail. They are mostly brown in color, and are very quick runners, both in the water and on land.

Second: Drakes are all bred in the water and are of various sizes and colors, abounding in all their varieties in vast numbers, from the large green drake to the very tiny white drake. Their movements are sluggish on land; they will even allow themselves to be taken up by the wings. They are not so hardy as the browns. Their shoulders and bodies are exposed; but nature has furnished them with a temporary covering, which they cast off when the weather suits, bursting open the covering at the shoulders and coming out a different color. They have close, thick shoulders, and smooth, tapering bodies which curve upward like the feathers in the tail of a drake. They have a pair of smooth, oblong wings, which, when at rest, stand upright close together; a small wing stands at the root of each large one, and there are two or three hairs in the tail. Some species hatch out in two or three weeks. Others continue hatching through the entire summer.

Third: The duns have two long feelers, small heads, short necks, and small, jumped-up shoulders. They have two pairs of large wings, set near the head; the under wings of some fold double, and all lie close together along the back and slope down over the sides, growing broader at the ends. Duns all breed in the water from creepers that are enclosed in artificial cases, ingeniously fashioned around them for self-preservation. Trout consume these creepers while in their cases at the bottom, also while they are rising from the mud to the surface, which they do twice a year. When on land or at the water's surface in repose, they are similar in appearance to moths. They sport on the wing more in the dusk and twilight than in the daytime, and in general they are tender and susceptible to cold. They are more numerous on warm evenings, flying in quick whirls. When they alight they run exceedingly fast.

Fourth: The class of spinners, the long-legged, slender tribe of insects very alluring to trout. Some are bred on the land and some in the water. They have in general two feelers and a small trunk at the nose; six long, thin legs; and a pair of long, narrow, transparent wings—some slanting upward from the shoulders, others lying horizontally on each other over the back. There are vast varieties of them, from the large gray spinner to the small mosquito.

Fifth: The house-fly. The flies of this class resemble the house-fly in having large heads, thick shoulders, a thick body which is about half the length of the fly, and a pair of clear, oblong wings which lie flat or horizontal and point more or less from the body. They are principally bred on land;

but, being exceedingly numerous, many fall casually onto the water on windy days and are taken by the fish. The cowdung and the bluebottle are well known examples of this class.

Sixth: The beetles. The outer parts of these insects are hard and shelly, and the shoulders are united to the body by a flexible joint, which enables them to turn and steer; the two fore legs are attached to the shoulder, and the other four to the shelly breast-plate. They have two pairs of wings; the upper ones, which are hard, stand close to the shoulder-plate and fold over a pair of soft ones and the upper parts of the body. They are of an oblong or oval shape, more or less flattened. Many of them are bred in the water, and are very natural food for fish. They are a numerous class.

Seventh: The class of ants consists of many species that live in communities, often of immense numbers, and are dispersed over the fields and woods in places of their own peculiar choice. A portion of each community are annually furnished with wings; and in the summer season, at their appointed time, these fly off and leave the colony, as bees do their hives. Great numbers of them fly over and fall onto the neighboring streams and are readily nipped up by the fish. The working portion of the community have large hawk-like heads, and large oval bodies—which are united by two or three comparatively very small shoulder-joints, and to these their legs are attached; but those with wings appear to have but one jumped-up shoulder, as thick as their bodies, and united by a small, hair-like link. They have two jointed feelers which they use constantly; and they run exceedingly fast. Some species have one pair, and some two pairs, of thin, glassy wings, which fall flat over the back and reach beyond the end of the body. Their colors vary from black to green. The sage-green ant observed in August in such vast numbers I do not find mentioned in English books; their red ant I have not observed on our streams. The large black ant has a flight in May and a lesser one in June.

The first business of the angler when he arrives at the stream is with the aquatic flies of the day. If he cannot see these out on the water, he may often discover them on a spider's web, or he may find them with their creepers at their times of hatching, at the edges of the stream. An hour or two spent in research and observation, at intervals through a season, will give a truer and more correct knowledge of the right fly to use than many years of angling; and it is often the shortest way to get fine sport.

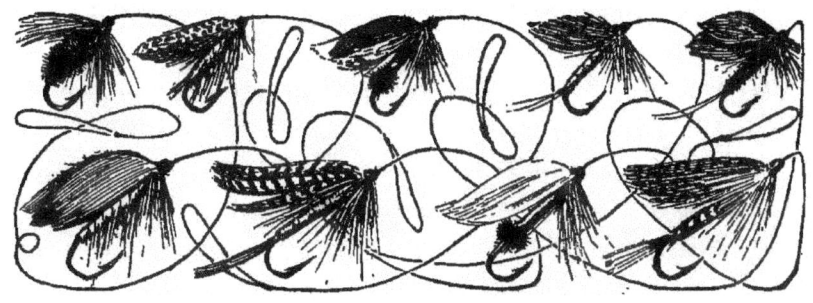

III

THE ARTIFICIAL FLY OF COMMERCE

It is incomprehensible why British and American fly-makers in recent years have gone out of their way to tie "fancy" flies when the natural insects are so beautiful in form and color; so varied, so graceful, that if they are copied true, fancy flies are exceedingly commonplace when they are put side by side. It is obvious that the originators never saw, or considered, God's handiwork: perhaps, in their egotism, they tried to improve on it; but it is certain they miserably failed, as we all must do, in any like attempt.

We have become so accustomed to using certain well-known artificials with more or less success that our confidence in them has become inflated, or at least satisfying.

Mary Orvis Marbury gives some amusingly naïve accounts in her book, "Favorite Flies," of how some of these historic masterpieces were born.

Most often some well-known angler takes up an old favorite fly, ties to it a bit of red or white feather, makes a test under most favorable conditions, and then swears the additional ornament was surely the cause. What is the outcome? He cries out to all his friends of the wonderful discovery of a new killing fly, then modestly names the new creation after himself or the place he happens to be fishing. Rube Wood, Seth Green, Cahill, Ferguson, and others are examples of them. By this I do not take from their usefulness or value as flies, but merely comment upon their rise into fame, as flies.

Tradition has much to do with such glaring mistakes. Each generation of anglers go on using tackle-shop flies without thinking—or, rather, what little time can be spared must be devoted to the actual sport of angling. They accept the wise and serious air of the tackle-dealer's clerk; swallow his assurance that such and such a fly is a deadly killer, when ninety-nine per cent. never caught or even saw a trout except the stuffed images or the wretched pictures supposed to depict them adorning the walls of every up-to-date store.

Another good reason why so many of the trade flies should be discarded is that for cheapness and profit anything will suffice if the hook is covered and a gay pair of wings attached in the same old way. To supply the varied wants of his patrons

THE ARTIFICIAL FLY OF COMMERCE

the tackle-dealer's stock must be enormous. You will naturally ask, "Then why add more?" My object is not to increase them, but to simmer them down to nature's reasonable limit.

Experts have told me that their list of flies for the entire season is confined to twelve varieties. I should hardly consider such a drastic cut-down sufficient to get fish or to enjoy good sport. I do think a careful selection of six for each month would be a better outfit to attain the result we desire— fair sport and fair fish.

May is the only month of the entire season when trout are not particular which fly they take, and the reason for that is quite plain. Natural insects are then so abundant and so varied that all are alike to them. After the winter's comparative fast, trout are unusually ravenous. When the season advances you must needs use more judgment in the kind of lure you offer, or you do not get a rise.

Of course most anglers have one or more favorite flies that reminiscence of happy days and battles won make it hard, from a sentimental point of view, to throw aside and forget. But I maintain that we fish for pleasure and success. Memory of the past is all very well; but hope is the fisherman's guide to beat our own and our brother's record. That success may be achieved in a higher, more learned way by taking nature into our confidence and using flies that do resemble, as far as possible in wool and feathers, the insect trout take day by day as food.

I was asked by the author of "Dry Fly in Swift Water" whether I imagined I saw the natural insect as the trout sees it? My response was, "Most certainly, when looking at its under body." The fine distinction of transparency is, in my opinion, too far-fetched; and so, too, is whether trout see colors as we do. It seems impossible to conceive that a trout will pause to consider if an insect is sweet or sour, bright or dull, large or small. Were I a trout, I would be quick to seize the fat, juicy body of the brown drake, in preference to the hard, small body of the black ant. But, if no drakes were rising, ants would suffice.

This will furnish an example of the wide difference in opinions from the supercritical dry fly exponent to the general fisherman who takes for granted existing conditions as he finds them.

If the angler will carefully study a page or two of colored commercial flies, he will observe a continued weariness of the same cocked wings spread outward, the same shaped body, all of the same size; the only difference being in the color of wings, body, and hackle. Let them be compared with any one of the monthly plates in this volume—no matter if it be the natural insect or the imitation flies. See how varied nature is, not only in the classes, but in

THE ARTIFICIAL FLY OF COMMERCE

the remarkable variety of different members in each class.

Aside from color, if you take the family of drakes, for example: Every part of the body, head, tail, wings, and feet of each insect is distinctly different from the other in form and size. Nature never repeats itself. When properly classified there is a gradual development of form and color; gradual, it is true, but very decided in its infinite variety of beauty.

While wading a trout stream you will observe that nature is always varied: large flies and small flies; when they alight their wings are cocked, or sloping over the body, or they may lie flat on the top. You may depend upon it that trout observe them, and know it just as we do; for all through their life they are on the watch for food.

IV

TROUT FLIES IN APRIL—WHEN INSECTS FIRST APPEAR

In April "Sol wins the ascendency, and blunts the sharp teeth of rebellious winds—withered winter vanishes in flowery green and woodland music."

The native song-sparrow carols merrily; the bluebird flits by in familiar companionship with the angler alongside the stream. The fish-hawk has returned from warmer climes—bolder, because more hungry in spring than in summer; the kingfisher, robed in gray, attends to the business of depleting the stream of young fish.

"Trout are now voracious and bold. They dash unerringly at the passing fly without fear or scruple. This genial life-cheering month teems with sport for the fly-fisher."

Suddenly, any day, all along the river trout stir up as if a general had commanded them to begin feeding; but not before the waters run much lower and the refreshing warm showers come; then "flies follow flies in thick succession fast."

In early spring when trout begin to feed they are

in dead earnest; they live only to eat—much more so than animals and birds, whose habits require time to bathe, trim feathers and fur, and time to spend in building and preparing a home. No creature living on land has nothing to do but feed continuously day and night, as do trout during May and June. I have opened many fish, caught on the fly or with worms, that were stuffed full of food, yet ravenous for more.

Trout are forced to fast in winter because nature is dormant in the very cold months. Minnows and other small food fishes lie hidden under rocks or embedded in mud and sand; insect life is the same. But immediately the ice moves and floats away, the creepers stir about the river bed, and they furnish the very first food of trout at the beginning of the season.

The warm rays of the sun rapidly gain power during the first part of April, and the most hardy creepers begin to rise at the surface in sheltered spots. Small, dark-colored duns and some browns are the first to appear. It was not till the 28th of April, a fair, warm day, I observed the first brown drake (March brown), the largest and most important fly in April; a week later drakes were fairly plentiful, and trout were visibly feeding on them as they fearlessly sailed like little brown yachts down the stream.

In the higher mountain altitudes you can safely

figure on ten out of the last fifteen days of April being cold and chilly, with heavy frost and snow flurries. The first fifteen days of May the nights are almost always frosty. In 1915 there was a heavy frost as late as June 13th.

Nos. 1 to 7 of the flies pictured on the colored page for April were captured from the middle to the last of the month when the water was very high and cold from the melted snow, which was still visible in drifts on the north side of the mountains. The season was abnormally late—two weeks at least. Snow was falling at intervals, but it melted as it fell. Consequently no creepers were visible on the stones at the side of the stream or in the water. Minnows, chub, and dace had not come up into the shallows. What trout I captured in the river were taken with worms; and they were all native speckled trout and rainbows; no brown trout responded to the lures. Upon cutting open two large fish there appeared a mass of black creepers, some of which were quite large and not yet seen at the riverside; showing conclusively that fish had begun to feed on river-bed creepers, but not as yet on surface flies. I could not possibly rise a fish on any sort of fly; the trout were (for the time being) absolutely quiet, lying in deep pools underneath the rushing water, waiting for warm, sunny days and higher temperature in the stream after the snow-water should flow away.

April Insect Chart

Numbers marked with asterisks are choice flies finely tied from the author's patterns and sold by his agents

No.	Name	Date of Rise	Time of Day	Weather	Family	Order
1	Needle-Tail	first to end of month	any time	cold, cloudy	Dun	Trichoptera
*2	Brown Buzz	first to end of month	any time	cold, cloudy	Dun	Trichoptera
3	Short-Tail	early to late	any time	cold, cloudy	Dun	Trichoptera
*4	Brown Drake	late, scarce	all day	warm, bright days	Drake	Ephemera
*5	Long-Tail	late	all day	cold, cloudy	Drake	Ephemera
6	Soldier Drake	middle	all day	warm days	Drake	Ephemera
7	Sailor Drake	middle	all day	warm and cold	Drake	Ephemera
*8	Redbug	early	afternoons	warm and cold	Beetle	Colesptera
*9	Longhorn	early	all day	cold days	Stone-fly	Perlidae
*10	Cowdung	early	all day	warm, windy days	Two-wing	Diptera
11	Needle Spinner	early	all day	warm days	Spinner	Diptera
12	Nobby Spinner	middle	afternoons	cold days	Spinner	Diptera
13	Little Mauve	early	all day	warm days	Dun	Trichoptera
14	Cocktail	middle	afternoons	cold days	Dun	Trichoptera
15	Blue Cocktail	early	all day	warm and cold	Dun	Trichoptera
*16	Male Shad-Fly	late	all day	warm and cold	Dun	Trichoptera
*17	Female Shad-Fly	late	all day	warm and cold	Dun	Trichoptera
*18	Female with Eggs	late	all day	warm and cold	Dun	Trichoptera

Note—*Shad flies also occur in May*

Chart Key to Enable Anglers to Fish Intelligently According to Time, Date and Rise

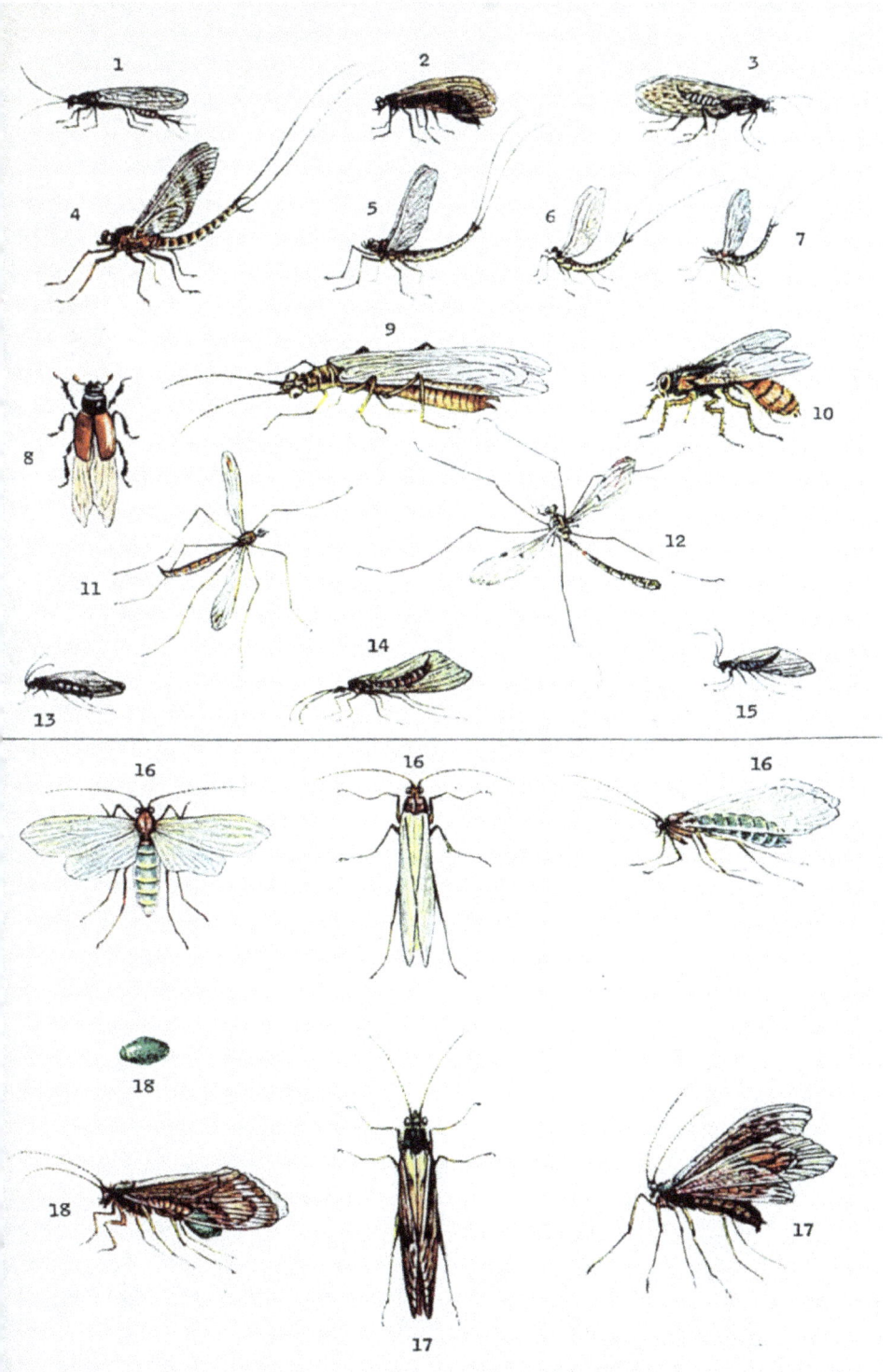

A SELECTION OF THE BEST TROUT INSECTS FOR THE MONTH OF APRIL PAINTED FROM LIFE BY THE AUTHOR

TROUT FLIES IN APRIL

From the Atlantic to the Pacific of the temperate zone the prevailing temperature at all seasons of the year is so entirely different from that in England that nothing in English fishing literature can be applied to our streams. Our mountain rivers and lakes freeze over to a depth of 16 to 20 inches in winter, to break up in ice gorges and floods in spring, while the summer heat and periodical droughts produce conditions upon aquatic insects and fish life that are inconceivable to the British angler.

The awakening of spring is erratic. We cannot always rely on days in April balmy enough—especially in mountain regions—to produce a rise of trout flies. But in sheltered, sunny places there are a few early, hardy flies flitting about in cold weather, for a few hours around midday. Most of these flies are small and dark in color. If you try a fair imitation, even in places where they are not on the wing, trout do sometimes rise to them; but not till the snow-water has run out. Later on, thousands of browns and duns which have left the creeper state and are lying sheltered under stones, waiting for the warmth, appear in flight quite suddenly after a couple of days' warm weather.

The first fish to rise to the artificial fly are the little redfins, and another day or two of fine weather the native speckled or rainbow trout may be seen feeding at the surface.

If you are fortunate enough to see a trout feeding on a certain species of fly, it is easy enough to imitate it and offer your fly to the fish. I always keep a sharp lookout on those insects most abundant over the water, and imitate them as nearly as possible.

Insects rapidly increase during the last few days in April, both in variety and numbers; so that when May comes in, the number is three or four times greater.

The more I study the subject, the more certain I am of the wisdom of confining my attention to and being thoroughly acquainted with those flies best calculated to kill trout, be they rainbows, speckled, or browns. Fortunately the same flies lure equally well all three species of trout.

It is natural to suppose that if trout do not respond to your flies they are feeding on creepers at the bottom, to rise occasionally when a tempting titbit floats by overhead. Trout undoubtedly must be feeding continually on these creepers, both on the bottom and while they are ascending to the surface for the change into the imago state, even before the short time it takes to form the wings perfect enough for flight. Thus vast numbers of them are devoured.

No. 1. Needle-tail dun. This active little antlike fly runs along rapidly, and just as quickly it wriggles over the water's surface. It is similar to what is known as the needle-brown and is always mentioned by anglers native to the stream as a black gnat. While in motion the two feelers are constantly moving back and forth. The dark-brownish body is quite plump, though hidden by the overlapping wings, which make it appear very much like the black gnat. Near the end of the tail are two short stylets; the tail, elongated from the body, has two shorter stylets growing out at the end; and the whole seems to drag on the ground or water as the insect moves along. On the day this was captured the weather was cloudy and the water very cold, yet they were quite abundant, both flying and resting on twigs or stones at the water's edge. I have proved it to be a most excellent fly if tied with plenty of hackle at the shoulders and tail, with dark ribbed body, and wings a steel brown. It is the first fly I should use very early in the month on cold, cloudy days.

No. 2. Brown buzz. This is a small, dark dun, with a black body like the ant, but larger and stouter. The dark-brown wings should lap well over the body. A curious feature is the absence of either feelers or stylets. It was caught while in flight, numbers of them being congregated buzzing up and down near the edge of the stream. The local (Catskill) name is brown gnat. It disappears in warm weather.

No 3. Short-tail. I caught this flying over the

water on a bitter cold day when it was snowing. It has unusually large shoulders and a very small, black body, with wings that appear yellow while in flight. The gray-yellow wings are mottled in brown, and the black legs are quite hairy. This is another insect of larger size in April. I am confident of its ability to lure trout better than any existing pattern for cold and stormy days. My imitation has been tested and found good even during cold days in May.

This fly was absent the latter part of the month when the temperature was much higher. It is of less consequence what date these insects were captured; but it is very important to know what kind of weather they are on the wing. A study of the monthly chart will enable the angler to know what artificial he should use in different weather conditions.

No. 4. Brown drake (March brown). This has its greatest rise in May, but if the weather is warm at the latter end of April it appears in flight in fair numbers. The first specimen I saw was on the 28th, in balmy springlike weather. I shall defer a description of this insect till May; but as it sometimes does happen that warm weather comes early, it would be advisable to have a few brown drakes in stock ready for use, because this fly is of the first rank in getting a rise of trout.

No. 5. Long-tail drakes. So named from its

tail being longer than that of the average drake. It appears in flight rather late in the month and may be seen on cloudy days over the water, rising up and down in quick movements, along with spinners. It is an excellent fly, that may be used at the opening of the season till the middle of May, and later if May is cold.

No. 6. Soldier drake. So named from the bright brown-red coat on the head and shoulders and the yellow under body. It is a lively little fly which first appears about the middle of the month if the sun is out. It appears all day if the weather is fairly warm; but on cold days after emerging from the water it creeps under stones at the water's edge to be protected from the cold. It is a very good small fly. Its beautiful, fat body is formed in a graceful curve as it stands proudly upright on its tall legs, and the long stylets curve from the tail high and elegant above the water. The wings are a creamy white, and the under body and tail a soft, pale yellow.

No. 7. Sailor drake. This appears at the same time as the preceding, but is more hardy and is present on the water on both warm and cold days. It is a very small, yet attractive, dark-bluish drake, most often seen in bunches or groups flitting about near the water in sheltered places on cold days and out on the river when the sun is bright. At intervals some leave the group near or under the bare branches of willows. It is a pretty little fly, dull in color; and it is one that cannot be left out of an April list of good flies.

Ronalds describes two flies, the "jenny spinner" and the "iron blue dun," which seem to correspond with these last two flies, the soldier drake and the sailor drake; the sailor being captured six days after the soldier. He says: "After two days the iron blue changes to the jenny spinner." I can furnish no proof that the sailor drake was transformed from the soldier drake, though I captured many specimens both at rest and on the wing. I should, however, much prefer to have them named in their class as small blue and brown drakes, instead of duns and spinners.

No. 8. Redbug. This handsome little beetle is exceedingly plentiful early in April on warm days. It flies swiftly and is seen a great deal over the water's surface; though on land it prefers sandy places, from which I imagine it breeds from a land creeper. In flight its appearance is very attractive, very much like the English artificial fly named "marlow buzz." I saw several specimens floating on the surface with wings outstretched; and I should think that two imitations, one at rest, the other in flight as seen skimming close to the surface, would be an excellent April fly. The one I have tied, copying the insect at rest, has been tested, and trout take it very well indeed in the afternoons of warm days.

Locally, it is styled the redbug; and native anglers try the Montreal fancy fly as an imitation of it.

No. 9. Longhorn. This one I name longhorn; it is locally misnamed the willow fly—so called, I assume, from its habit of resting on and walking rapidly along the pussy-willow stems. It is truly of the "brown" or stone-fly class and very abundant. When just hatched its body is a bright shiny orange, with four glossy gray wings overlapping the body. I saw many specimens of various sizes on the willows when the sun was out, but on cold days they all creep under the stones to shelter from the snow and wind. Later, as the weather gets warmer, the flight is very thick all day long and at evening till after dark. They are fat and must be satisfying trout food both as creeper and fly. I have made the artificial of deep orange on the body, with flat wings, and it has proved one of the very best early flies. A similar fly appears later, in May, but in this one the stylets are absent, the head is smaller and more pointed, and the eyes are bulging.

No. 10. Cowdung. A well-known and most excellent fly, when tied properly. It appears quite early on bright warm days, and the artificial is best used on windy days. A more detailed description of this insect will be given in May, when it is much more abundant.

No. 11. Needle spinner. This flies around early, along with various duns and small drakes, in the daytime on warm days. It is quite plentiful, congregating in bunches six feet above the water's surface, to drop down now and then and also to ascend higher. It will prove effective when lightly dropped onto the surface and floated along a runway.

No. 12. Nobby spinner. This is larger than the previous insect, and appears later in the month, on warm days, mostly afternoons. When the sun is absent it finds a sheltered spot, to be protected from the cold.

No. 13. Little Mauve. An early, dark-colored dun, commonly called a black gnat—indeed, the natives style a great many insects of the dun class the black gnat. It is an excellent small fly, and the imitation may be used all month.

In fact, all these April duns are useful nearly every month of the season, being so similar to the insects that rise later in warm weather; for that reason a supply should always be kept in stock: when the larger drakes are absent the dark, small duns induce trout to rise.

No. 14. Cocktail dun. A somewhat larger, brownish dun, with two long feelers and long hind

TROUT FLIES IN APRIL

legs. In flight its dark-greenish wings make it less conspicuous near the surface; but these duns are often very thick over the water on the afternoons of warm days. The flight becomes more numerous in May; so that the artificial may be used from the middle of April till nearly the middle of June.

No. 15. Blue cocktail. This is another variety of the dun class. It is small, but seems hardy and comes out in the coldest weather in both April and May.

These various little duns are most valuable, both in the cold weather of spring and in the hot weather of summer. They disappear in warm weather, but are so much like the summer gnats that the imitations suffice for both. The blue cocktail shown here was caught in normal weather, though the water was still ice-cold, and trout refused to rise.

THE SHAD-FLY

In consequence of the large number of insects that appear in May, I found it impossible to show on the May page of colored insects the various positions of the shad-fly I wished to depict. As the April insects are correspondingly small, I place the shad-fly on that page. Especially so, because I observed a few females (though small in size) on the water during the latter part of April. The female, without the egg-sack, can be used in any weather all day in the last week of April.

The shad-fly, during the great flight, is the most abundant trout insect food that appears during the entire season, on our Eastern, and some Middle and far Western, streams. For that reason it truly merits a more detailed description than any other trout insect. Trout are always ravenous for it, yet a true imitation of this handsome fly has never been made.

Numerous and varied are the reasons for its name of shad-fly. My old friend, William Keener, a famous fly-fisherman of Roscoe, N. Y., says this typical aquatic insect has been known as the shad-fly for seventy years at least, and it is so named because its flight occurs just at the time when the shad arrive at the headwaters of the Delaware River to spawn; while so doing they feed upon this fly, thereby attaining a fat and sleek appearance after spawning time is over.

A second reason for the name is that the flight occurs exactly when the beautiful white woodland blossom known as "shad-blow" sheds its white petals, to be blown by the wind on the water's surface, lightly floating downstream.

Thirdly, the name was given because the egg-sack attached to the body of the female is very much like shad roe in shape and appearance, except that the color is green.

Of course, the shad-fly appears on rivers where no shad spawn or shad-blow trees grow.

TROUT FLIES IN APRIL

Both male and female are very beautiful in shape and color. The difference, especially in color, between male and female is greater than in any other aquatic insect I know. From a glance at the colored representations it will be seen that the male has two large and two small wings of silvery transparency, tinged with warm yellow, that lap over the body like most species of duns, although they rise higher over the body than the wings of the female. The tail, in seven segments, is a beautiful, soft, gray-green color, in striking contrast to the vienna-brown shoulders and long, pliable legs. The head is small, with shining black eyes, between which grow two brown feelers or horns, moved forward or backward at will.

The more sedately colored female has four gray-brown, mottled wings that lap down just over the tail. The shoulders are bronze-green above, turning to purplish-black at the belly; and the feet, as well as the horns, are brown. The black head and eyes of the female are somewhat larger than those of the male. The tail is dark brown, with a dull band of yellow ochre along the sides. At the end of the tail, pushed in a sort of cavity, is the bright green egg-sack, which is easily removed. I pulled quite a few egg-sacks from the body and placed them in shallow water; after a few seconds they sank to the bottom. I have so far not been able to tell for certain whether the shad-fly deposits the egg-sack on the water's surface, where it frequently alights, or drops it while in flight; I should incline to believe in the former.

Both the fly and the egg-sack vary in size. Large females measure fully half an inch from head to end of egg-sack, and three-quarters of an inch from head to tip of wings. Egg-sacks are one-eighth of an inch, more or less.

It will be seen that the egg-sack in trout diet is a juicy morsel, equal in food quantity to, if not greater than, many of the bodies of the small drakes and duns.

I have been more particular than usual in the description of the shad-fly, because, while the great flight is only two to four days, the insect rises (more or less) during April, May and June; and my artificial representations of both male and female have succeeded in rising trout in a surprising degree in all those three months.

In consequence of my absence from the river one season during the annual rise (which was the smallest on record) I had to wait a year to get the female with eggs. Then, however, I was most fortunate to witness what I may term a preliminary rise—that is, a first great hatch, which generally appears before the real or final hatch, when the vast clouds float along over the river for a few hours like a severe snowstorm reaching up both forks of

TROUT FLIES IN APRIL

the river (Beaverkill and Willowemoc) to a distance of over twenty miles.

The early part of May, 1915, up to the 19th, was unusually cold and stormy, with frost almost every night and vegetation two weeks late. On the 11th and 12th of the month the temperature suddenly changed to almost summer heat. This brought out a fair rise of shad-fly, mixed with a heavy flight of small, iron-colored drakes and yellow sallies. Then came a sudden decline in temperature to bitter cold. Two days afterward, with slight warm showers and rising temperature, the great rise began about noon, the insects flying thick from the surface to thirty feet in the air, and all remaining over the water till sunset. It would be a feast to the eyes of most anglers to see the water's surface fairly bubble with rising fish, fighting-mad to eat.

The shad-fly is exceedingly active and strong in flight, which makes it most difficult to capture. It is somewhat like the house moth in action, fluttering, rapidly moving in all directions, to get out of the way of a fly-net. It is utterly impossible to capture a specimen away from the water; as, on very rare occasions when they do fly over land, they go from twenty to forty feet high. Sometimes, but not often, they do take a rest to alight on leaves and twigs at the water's edge, but their movements are rapid in every way when efforts are made to capture them. Yet in flight they appear to move slowly and quietly along.

The shad-fly is an excellent floater, and spends much time walking along the surface, which it does with ease.

While most trout flies hide themselves behind large stones and under leaves for protection from the wind and rain on cold days, a remarkable feature of the shad-fly is to gather together in a great swarm on a rock by the riverside, something like a swarm of bees. If you brush them off in large clusters to the water's surface they spread out, floating down with the current, to attract a surprising number of trout and chub which rise up, making the water bubble in all directions, to feed and gorge on the unexpected feast.

I can with confidence predict that, for the latter half of May and most of the month of June, the artificial copy of this insect, either male or female, will be found by fly-fishers (both wet and dry) to be superior to any other fly, even including the large green and gray drake. When the shad-fly is on the water you never fail to see trout rising; and when the great rise appears, it is impossible to catch a trout with any prevailing fly as now tied.

V

TROUT INSECTS FOR MAY

THE smiling month of May is indeed the angler's halcyon days. Hungry trout are in abundance; they forage and feed without fear or scruple. When the weather is warm myriads of flies flock the air and the trout revel in food night and day.

Yet it is not always so. May is a saucy, coy month, changeable as the wind, from good to bad, and bad to good, all of a sudden in the mountain regions, during the first week or two. After a beautiful day, you may wake up the next morning to see a fierce snowstorm or perhaps a heavy frost. Toward the end May robes herself in bright emerald, and the woods and riverside are spangled with spring blossoms and golden pussy-willows. I can imagine nothing more delightful than fishing for trout at the end of May.

In 1915, during the early part of the month the sky was cloudless, the weather warm though windy, and perhaps a little in advance of the average May. There was no rain or frost the first week, and the river for so early in the season was somewhat low. The coy month enticed vegetation to spring forth, and to the insects it proved a gay deceiver; for during the latter part such abnormal conditions prevailed as to change its smiling appearance to frigid February or to weeping and blustering March.

New York's able and efficient State Fish Culturist, Dr. Tarleton Bean, had allotted the Beaverkill the year previous a good stock of fry, part of which were rainbows—now nearly two years old. I anticipated good results; and every angler on the stream was fully satisfied.

I rarely catch a brown trout during the latter half of April; and when I do, they are poor, thin, ill-conditioned things. In the Catskill region the brown trout begin heavy feeding about the middle of May, and they soon regain their plumpness by continuously eating, night and day. At this season they never stop eating, as they do in summer, when they feed only at night from sundown.

On the first of May I caught a nineteen-inch trout on a big nightwalker during a sudden stream-flood. It was so emaciated and thin as to weigh but a pound and a half, whereas in fine condition it would have been at least as heavy again. As I would never eat such a skinny-looking thing, I put it back in the stream, uninjured. With speckled trout conditions are entirely different. They are exceedingly active in play when captured, attacking

May Insect Chart

Numbers marked with asterisks are choice flies finely tied from the author's patterns and sold by his agents

No.	Name	Date of Rise	Time of Day	Weather	Family	Order
*1	Green Drake	late	any time	warm days evenings	Drake	Ephemera
*2	Brown Drake	early to late	all day	warm days evenings	Drake	Ephemera
3	Black Ant	middle	all day	warm days	Ants	Hymenoptera
*4	Mottled Drake	middle	all day	warm days	Ants	Hymenoptera
*5	Cinnamon	early	evenings afternoons	cloudy dark days	Dun	Trichoptera
*6	Sandy	late	evenings & mornings	cold days	Dun	Trichoptera
7	Purple Drake	early	mornings	cold days	Drake	Ephemera
*8	Gray Drake	middle	all day	warm days	Drake	Ephemera
9	Gauze-Wing	middle	afternoons evenings	warm days	Stone-Fly	Perlidae
*10	Yellow Sally	early to late	evenings	warm days	Stone-Fly	Perlidae
*11	Flathead	early	all day	warm days	Stone-Fly	Perlidae
*12	Alder-Fly	early	evenings	all times	Dun	Trichoptera
13	Orange Stone	early to late	all day	all times	Stone-Fly	Perlidae
14	Horned Spinner	early to late	all day	all times	Spinner	Diptera
15	Crane Spinner	early to late	all day	all times	Spinner	Diptera
16	Green Spinner	early to late	all day	evenings	Spinner	Diptera
*17	Golden Spinner	middle	all day	warm days	Spinner	Diptera
18	Cotta-Fly	late	afternoons	warm days	Four-Wing	Hymenoptera
19	Glossy-Fly	middle	day and evenings	all times	Four-Wing	Hymenoptera
20	Bluebottle	early	all day	windy days	Two-Wing	Diptera
21	Cowdung	early	day time	cold days	Two-Wing	Diptera
22	Yellow-Horn	early to late	evenings	all times	Dun	Trichoptera
23	Speckled Dun	early to late	all day	warm days	Dun	Trichoptera

Chart Key to Enable Anglers to Fish Intelligently According to Time, Date and Rise

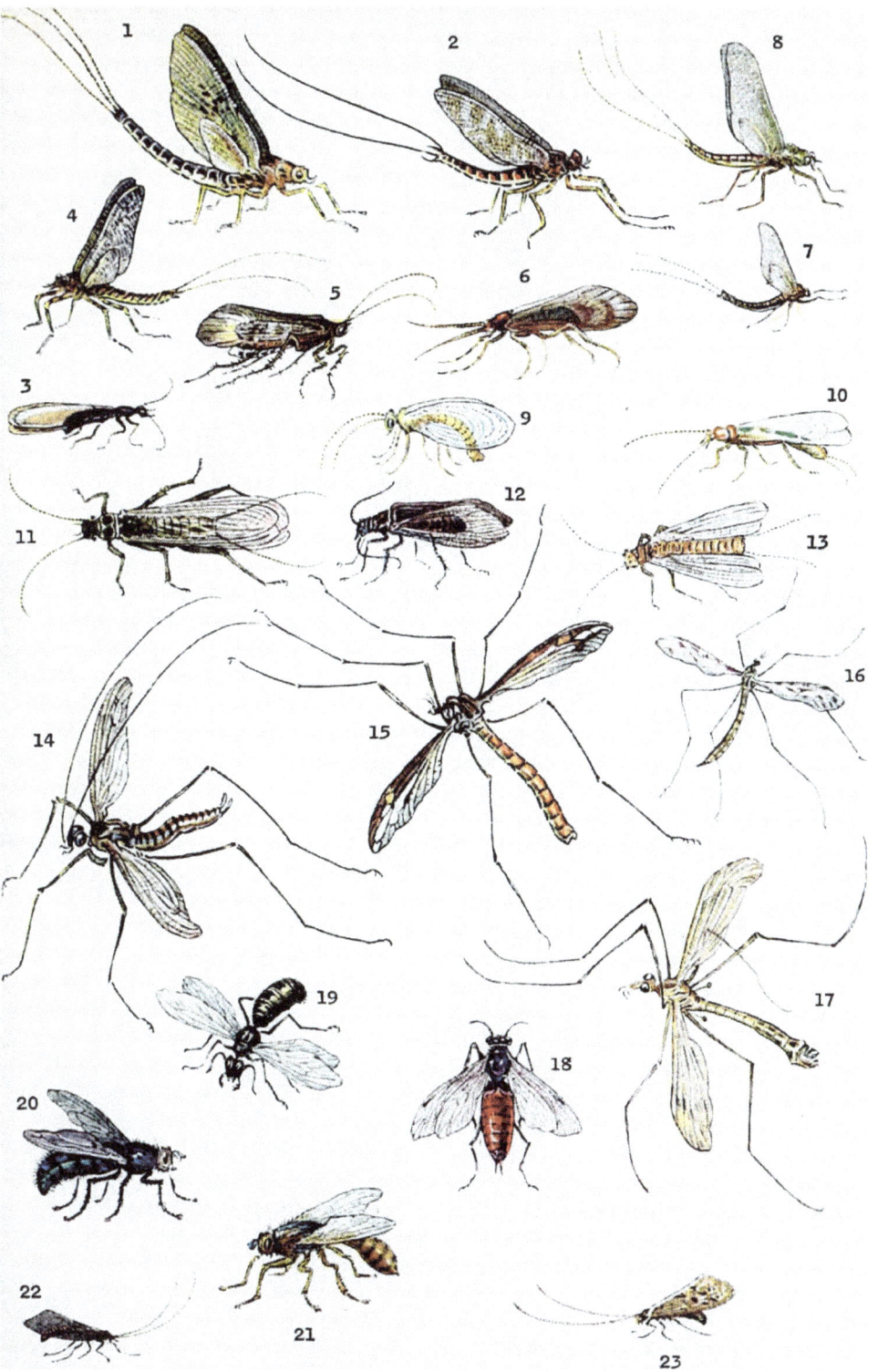

A SELECTION OF THE BEST TROUT INSECTS FOR THE MONTH OF MAY PAINTED FROM LIFE BY THE AUTHOR

TROUT INSECTS FOR MAY

both fly and bait with equal vim, and are plump and in splendid condition when the season opens. This establishes a fact, I think: that the brown trout does little or no feeding during the winter, and that the brook or speckled trout feed at all seasons. I have opened many a well-conditioned native trout early in the season, to find that their stomachs contain very little food—mostly small creepers from the river bed, a mixed mass, black in color, showing they devoured both creeper and case. Our native trout, I think, does not begin to eat fish food till it attains a fair growth of about ten inches in length, whereas the brown trout takes very small fish as food when only five inches long, or about a year old; they will take artificial flies before that.

Many of the insects captured in the first few days of May were the same that I observed late in April. A fairly good rise of yellow sallies appeared the third day; and the brown drake came out in increasing numbers. The large green drake did not appear till the 18th. Every day the shadfly became more numerous; so, too, the cowdung and the bluebottle were quite plentiful.

No. 1. Green drake. This is the largest (except the big stone-fly), though not the most beautiful, aquatic insect that trout feed upon. Its long, fat body proves a very alluring bait. Indeed, it is so good that even very poor imitations are greedily taken by large and small trout during the entire period that the natural insect moves over the water.

The green drakes do not appear to rise over the waters of the Catskill region in vast clouds such as are described on British streams, or some other American waters, as, for instance, Lake George. They are, nevertheless, quite abundant on the Beaverkill and neighboring streams, both in the daytime and at evening. I have counted forty specimens of both sexes in one locality floating and flying over the water.

The female floats gracefully along the surface for a considerable distance, at intervals rising and dropping until devoured. In its flight it is of a decidedly yellow-greenish tint, and it lives three or four days as here pictured; then the female changes to what is known as the gray drake, casting away its garb of delicate yellow-green and appearing in one of soft gray. The wings become more transparent and sparkling, and the fly more active in this, its final, existence. The male, smaller than the female and not nearly so beautiful or so fat, changes to what is known as the black drake. The under part of both male and female is pale yellow.

The green drake, while the most luscious and tempting of trout food during its short rise in the month of May, is not by any means the most abundant of the insects found in the locality here described; though a good imitation will be found an unfailing lure for the large-sized brown or rainbow trout. This splendid fly, alternated with the brown drake, both tied on No. 6 or No. 8 hooks, would be unequaled for use in the last week in May and the first week in June. However good this fly is known far and wide to be, personally I should give it second place to the female shad-fly when that insect is rising and laying its eggs.

No. 2. Brown drake (March brown). This has always been a prime favorite with both angler and trout. As for myself, whenever I see a cloud of these drakes joyfully flitting over the water, I put away other flies, and fish more earnestly than ever with the English dry fly imitation. The imitation of this insect by English fly-makers seems to me better than that made by American fly-makers; but neither imitation is true to nature in representing the under body dark when the insect is a light pale yellow underneath. The tail, also, should be cocked up, a detached body above the hook, and the stylets of greater length.

While isolated specimens of the brown drake are seen toward the end of April, in the warmer weather of May these insects gradually become more and more abundant on the Beaverkill, flitting up and down over the water, sometimes fifteen feet in the air, where occasionally the sexes meet and fall together to the surface of the water, when they are greedily taken up by the trout. The pliable, long, fat body of this insect soon fills up what space is left in the stomach of a trout. It is an excellent floater on the surface, and with its wings erect and close together it tempts a trout to feed as no other insect can. Toward the end of the month the rise gradually diminishes; but, as in April, so in part of June, they still are seen, though only in scattered numbers.

This insect is just about the right size for the average fish; rainbows are particularly partial to it, either in the daytime or at evening. In fact, it is the best fly of the month, taking it altogether, because no matter what the weather may be, wet or dry, warm or cold, high or low water, the brown drake will rise a fish.

No. 3. Yellow drake. While smaller in size, this is an excellent fly for all occasions during the entire month. Its yellow body is fat and must prove a dainty trout morsel; for, floating fearlessly, always at the surface, it is taken in great numbers. It is one of the few drakes with three stylets; the majority have only two. The rise begins early in the month, and later you see the yellow drake everywhere, both day and evening.

No. 4. Mottled drake. This handsome little fly is an excellent floater and for that reason should, when possible, be fished at the surface, where it may be seen plentifully at all times of the day and evening. The high, mottled legs raise the body to make it seem a much larger insect than it is. At evening, when mottled drakes are most numerous, congregating in bunches, then dropping to the surface, where they stay longer than is usual with other drakes, you see the trout taking them in large numbers.

No. 5. Cinnamon. The only specimen of this species caught during the month, though there are quite a number, both larger and smaller in size, of this caddis family. This particular insect, which floats on the surface quite frequently, may be tied to use by the dry fly method, casting upstream on rather rough rapids.

No. 6. Sandy. This little sand-fly is a trimly built insect that flits about rocks and pebbles, very active on its feet as well as on the wing. It is quite abundant and evidently a favorite trout titbit. Its four dark brownish-gray wings fold tightly over the fat, dark-brown body. At times, especially when in flight, the two horns, or feelers, are thrown backward over the shoulders in line with the body. Though quite at home floating on the surface, even in rough water, the sandy is nevertheless most frequently seen about the stony banks of the stream, at all hours of the day except from late afternoon till dusk, when it stays over the water all the time. It is then—late evenings—a most effective lure.

The English sand-fly is a fairly good representation of this insect. It does not appear to be popular with American anglers; though I feel sure that if the artificial I have tied is used it will prove a second-best to the brown drake. To be most effective it should be made to float; indeed, all insects that float naturally a good deal on the water should be so imitated and fished by the dry fly method. Those insects which do not and cannot float but are blown about the water and drowned, should have the artificial flies made to fish by the wet fly method.

No. 7. Purple drake. A small insect, which is very common all through May, and also in June. It rises in large numbers early in the month and may be seen on the coldest days. It is quite small and appears in flight much darker than on close observation. It is best used in early mornings before larger and more important flies are on the wing.

No. 8. Black ant. I fail to see why trout should condescend to take this insect, with its puny body encased in a hard black shell, when fat, juicy insects are almost always at hand. Yet, time and again, I have proved the artificial black ant to be excellent in coaxing fish to rise at all times of day. My own imitation was used in June and July with good results. In the middle and latter part of May the black ants swarm in large numbers, flying along the river in vast clouds. There are two sizes, both alike in shape and color. I have not yet seen what is known as the red ant; but in the plate for August

TROUT INSECTS FOR MAY

will be noticed the sage-green ant, a much larger and more desirable insect, which is described in its proper place. When at rest, the long wings of the black ant lie flat on the body and are of a glossy bronze color.

No. 9. Gauze-wing. As a choice titbit for trout, the gauze-wing is similar to the yellow sally. Its wings, which are of a silvery transparency, are wider and lap together more closely on each side of its long, greenish-yellow body. I have made the imitation wings nearly white, and have found it a good evening fly. The gauze-wing is very plentiful shortly before and after sunset. It flies very slowly near the surface of the water, where it alights at short intervals.

There is a beautiful representation of this insect in Ronalds' "Entomology," the wings being made of a pale blue dun hackle; but I should imagine that from the trout's point of view, a light wing over the green body would be truer to nature.

No. 10. Yellow sally. This insect, the first that was captured, is fairly plentiful early in May. Although a day fly, with a preference for dark days, it is most abundant at evening, when it may be seen moving slowly along the riverside, its fat, heavy body hanging down, as if too great a load were carried by its delicate wings. Sometimes in crossing the stream it is blown to the surface of the water, where it struggles ineffectually, unable to regain its flight. It is a solitary insect, never traveling in pairs or in groups. It is taken with great avidity by trout; though I have never had a rise on the artificial fly as now tied. The imitations all show bright yellow wings, erect or cocked, whereas the wings should be light blue-gray and lie nearly flat over the body, with the hackle, horns and tail of a deep yellow.

No. 11. Flathead. One of the numerous stoneflies seen in May. It is large enough to use in its natural state if empaled by the thorax on a No. 6 hook and carefully manipulated with a fine nine-foot leader, as you would a worm, in various runways or pools where trout lie. But I have had excellent results, in both June and July, from the imitations I have made in various sizes; as these flies appear to rise in goodly numbers every month of the season.

Stone flies vary in size and color, according to genus, and are found in greater abundance near the part of the river which has a stony or rocky bottom. When the creeper is ready to change into the insect, it seeks the edge of the stream and attaches itself by a glue-like substance to the under side of a stone. It then crawls to the upper side of the stone, or to the stem of an aquatic plant, where the skin splits open, permitting the winged or "perfect fly" to escape.

The stone-fly prefers to come out at dark, or on

dark days, and flies more often at evening. Its lower wings are much larger than the upper and are placed quite far down on the third segment of the thorax; thus in flight it appears like two insects moving very slowly in the air. When at rest the wings lie flat and hang folded together, a little wider than the back and extending some distance over the tail.

Stone-flies do not drop to the surface to deposit their eggs as do the drakes. Halford states in his "Entomology": "The female drops them probably while flying at some distance above the water, and they separate as they sink toward the bed of the river." The fly, when it first emerges from its case, is quite soft and of a pale yellow, but it soon changes to a deeper brown. At all times it is heavy in flight. The slightest touch disables it; and it cannot rise from the surface if, by accident, it touches the water.

No. 12. Alder-fly. While plentiful among the bushes and tall grass at evening, it does not seem inclined to hover over the water as do many other insects. Nevertheless, it is a favorite evening fly and should be used as a wet fly only. It is equally as good lure by day as by evening; though it is generally used as second fly, because while casting it cannot be seen so well as a light-winged fly. It is a very black insect; larger than the black gnat, its wings being wider and hanging lower down on each side of the body. When there is a dearth of flies on the water, it is a good fly to use as a test in getting trout to rise.

No. 13. Orange stone. A smaller and more brightly tinted stone-fly than No. 12, and with a few different characteristics, though in the main it is similar. I caught in May a great many specimens of this family, and only place this particular insect to show the more vivid color of the body. Its size is just right for trout, and my imitation is very successful.

No. 14. Horned spinner. (Slightly smaller than the representation.) The horned spinner is often seen among a cloud of brown drakes, where they fly over the water, dipping and rising as the two sexes meet. Sometimes the long horns, or feelers, project forward, then again lie flat on the back. In their wavy motions while flying they often touch the water, and are sure to entice a rising trout. When resting on a stone or leaf, their long legs fit quite close to the body and the horns lie backward.

No. 15. Crane spinner. This insect moves about, sometimes resting on large stones that rise above the water's surface, then alighting on smooth shallow water, where it is taken greedily by the big chub. Crane spinners invariably move in pairs; and, though not abundant, they may be seen at all times of day as well as of evening. The body of the insect makes a good-sized tempting meal for a trout; and a good imitation will not fail to get a rise. I do not often see this fine insect in places where trout lie in wait for food; it seems to prefer quiet shallows, or rocky boulders on the riverside where the water is not deep. On windy days large numbers are blown off the stones on to the water; but, unless quickly taken by the fish, they soon recover and rise from the surface. The June spinners, quite similar in form, are very different in their habit of frequenting the middle stream over rapid water.

No. 16. Green spinner. This little spinner is more often seen in rough water, and skims around near the surface. It is abundant all through the month, and during part of June.

No. 17. Golden spinner. First observed in the middle of the month, and not common till June and July. The general tone of creamy-orange makes it an attractive-looking fly on the wing.

Making artificials of these spinners is, so far as I know, an innovation, as I have not seen them tied to imitate the natural insect. I have tested the artificials with success.

No. 18. Cotta-fly. An insect fairly common at the riverside, where it flies about the willow and alder bushes. It was captured flying near the water's surface. The beautiful, deep terra cotta body is in striking contrast to the black thorax and head, with the wings of a gauzy blue. It makes a handsome artificial fly and can be used as a floater, fished over smooth places near the willows and alder bushes.

No. 19. Glossy-fly. An ugly-looking black insect, quite plentiful among the willows and flying about the alder leaves that overhang the water, hunting and feeding on smaller insects. The body is glossy black, with brown-gray wings. The imitation, as here shown, will doubtless prove a good lure if used as a floating fly near bushes; for the natural insect floats fairly well with wings cocked over the body.

On the Beaverkill, where this insect was captured, black and dark flies are greater favorites than flies with light wings and body, in the early season before the green and gray drakes are on the wing. This will be the case, no doubt, in other localities.

No. 20. Bluebottle. This represents the well-known bluebottle fly, of which there are many examples in various sizes. These insects are always more abundant near farms and dwellings or near refuse cast along the stream at the outside of villages. It makes a handsome artificial fly, and can be used as a floater or, with extra long hackle, played as a buzz fly, and fished over smooth places near the willow and alder bushes. Bluebottles are often driven to midstream on windy days, and, as they cannot float, are drowned. They kick frantically, spinning round and round till seen by a trout, when, *flop!* all is over. The artificial should be fished wet on windy days.

No. 21. Cowdung. This fly has always been a favorite with anglers. It is, during May, exceedingly plentiful over the water on bright, breezy days, and will drop and float for a short period; but it seems to be unable to rise freely from the surface like many of the purely aquatic insects. It makes an excellent floating dry fly; and the same may be said of it as a wet fly.

The cowdung is now tied in many different ways and colors, even showing the slight difference as to size and color between the male and female. Some fly-makers have even gone so far as to make a combination of the two. Trout are much too wise, when feeding, to have a preference for either sex of any insect except when females are egg swollen. Offer them a true imitation of an insect with which they are familiar, and ten to one it will be taken, no matter to which sex it supposedly belongs.

I have never seen either an American or an English imitation that showed the tail copied true. It should be much thicker at the end than near the thorax. There also should be five bands of green on the tawny orange tail.

No. 22. Yellow-horn. Another of the small duns, so very abundant during the entire month. The long yellow horns are used as feelers, being moved back and forth continuously. The dull gray-black wings lie close over the long, thin body. The imitation is useful at any time, and may be continued through June, as the yellow-horn continues to rise throughout that month.

No. 23. Speckled dun. This is very similar in size and shape, but much more attractively colored. It is out on warm days in great numbers; and the imitation makes one of the best small duns of the entire season. The speckled dun continues in flight through June, and a few may be seen in July; though the specimens I caught in that month were slightly larger and darker in color. It is a first-rate little floater; in fact, when the large drakes are absent these small duns are the flies best to use.

I venture to assert that this selection of insects is more than sufficient for the purpose of the practical fly-fisher during the month of May. Some of these insects continue over the water through June or even longer, while others appear before May—in fact, even as early as March. I also am quite sure these insects will be found useful for fishing in all the trout waters of the temperate regions, in the East and in the Middle West.

May insects are so very numerous that I found it the greatest difficulty to bring down, by elimination, the number to reasonable limits; I was guided in the selection by my experience in past years as

TROUT INSECTS FOR MAY

to which were most valuable in each class. I think it is desirable to cut down still further the number pictured on the May page to, say, eight specimens, viz.: two drakes, two duns, two browns, one spinner, and a two-winged fly, the cowdung preferred.

Each monthly page contains an average of twenty flies, and for the season makes a round hundred. I don't suppose any angler would want such a large variety, because it is always well to have more than one specimen of each fly in stock. It is for that reason I shall give a selection from each colored page of those flies mose needed, and best worthy to use.

It is impossible to tell in advance what natural insects are on the wing, because their flight depends so much on weather conditions; but I can, with some degree of certainty, tell the right times to use each of the different flies. And this information is given, as far as it is possible to do so, in the charts that accompany the colored pages.

VI

THE BEST TROUT INSECTS FOR JUNE

AFTER a glance at the page of insects for June it will be observed that drakes and spinners largely predominate over the duns. The duns are hatching very numerously, but the major part are quite small in size and not useful for imitation. In both April and May we see many more duns than drakes; but as the weather gets hotter the duns are still more scarce.

Many and varied are the drakes, as to both color and size, that are hatching and being taken by the fish in the morning hours, then again at evening, just before sunset, till dark, and throughout the night.

As the vital heat of the sun keeps increasing, so do the flies increase in abundance and variety. None is quite so large as the green drake of last month, but the average size is larger than the average for May.

Throughout May the fish are very ravenous; they become fat on the big stone-fly, green drake, shad-fly and other fair-sized insects; and thus it is

THE BEST TROUT INSECTS FOR JUNE

that sport is neither so continuous nor so ample in June fishing, especially after the middle of the month. It becomes necessary to use greater care in fishing, and to be more exact to imitate those insects now so abundant on the wing.

In June, trout take a noonday rest on hot days; you may now and then persuade them, but it is not so easy with bright sky and low water.

After the first week in June, 1915, all the large green drakes had vanished. The last shad-fly was gone by the 16th, to be replaced by extremely numerous drakes, of a fairly good size and varied in color and form.

A peculiar feature of most of the drakes is that the two fore legs are dark—generally the same color as the body—while the four hind legs are light in tone, mostly pale or bright yellow, and often mottled in brown. Another very important feature is that the under part of the body and tail is invariably light—either yellow, green, gray or pink—no matter how dark the top of the body and tail may be. This feature makes an exact imitation of very small flies much more difficult—to wind the lower body lighter than the top. The fly of commerce makes no effort in that direction. A still more strange feature, to be found only in the June insects, is that nearly all of the drakes have vivid green eyes; none of the April or May drakes have eyes of that color.

Four of the most beautiful June spinners are

here pictured. The big spinners pictured on the page of May flies were still present on placid waters, but these four were playing and were caught over rapid and turbulent water, where trout were constantly taking them under.

Spinners of various sizes are exceedingly numerous in June. Some go spinning round and round about six inches above the surface of the water at astonishing speed, playfully chasing each other for sexual purposes, never touching the water till they join, when together they drop onto the surface, float a short distance, only to be devoured by the trout. Others spin in wavy motions, constantly dipping, floating a while, then rising into the air—if not taken by the trout.

Most all members of the spinner class seem to be good floaters, and are very swift in flight. I lost two hours' good fishing in a vain effort to capture the dark, mottled spinner, though many specimens were all about me, as I stood in quite deep, rough water, where they stay, flying low all the time. They are not to be captured on land.

The beautiful golden spinner—also seen in May—is now all over the river, gracefully dipping at the surface, then ascending to the height of thirty feet in the air, where it floats around for a short time, to come down and repeat the dip.

Trout fishing in June can be very good and also very bad—according to conditions. The first two

June Insect Chart

Numbers marked with asterisks are choice flies finely tied from the author's patterns and sold by his agents

No.	Name	Date of Rise	Time of Day	Weather	Family	Order
*1	Female Green-Eye	first three weeks	day, best at evening	warm	Drake	Ephemera
*2	Male Green-Eye	first three weeks	day, best at evening	warm	Drake	Ephemera
3	Broadtail	early to late	all times	cold, windy days	Drake	Ephemera
*4	Greenback	early to late	evenings	warm days	Drake	Ephemera
*5	Yellow-Tip	middle	all times	any time	Drake	Ephemera
6	Spot-Wing	entire month	afternoons evenings	warm	Drake	Ephemera
*7	Lemon-Tail	entire month	afternoons evenings	warm	Drake	Ephemera
8	Shiny-Tail	early to middle	all times	cold and windy days	Drake	Ephemera
*9	Chocolate	entire month	late afternoons, eve.	warm days	Drake	Ephemera
10	Orange-Black	entire month	all times	any time	Drake	Ephemera
11	Tawny Drake	entire month	all times	any time	Drake	Ephemera
12	Blackhead	middle to late	afternoons evenings	any time	Drake	Ephemera
13	Big-Eye	middle to late	afternoons evenings	any time	Drake	Ephemera
*14	Pointed-Tail	early to late	all times	any time	Dun	Trichoptera
15	Goldrim	middle to late	afternoons	warm days	Four-Wing	Hymenoptera
*16	Emerald	early to late	afternoons evenings	warm days	Stone-Fly	Perlidae
17	Little Yellow Stone	entire month	all times	warm days	Stone-Fly	Perlidae
18	Black Dun	early to late	all times	warm days	Dun	Trichoptera
*19	Hairy Spinner	middle to late	evenings afternoons	warm days	Spinner	Diptera
*20	Goldbody Spinner	early to late	evenings afternoons	warm days	Spinner	Diptera
21	Tiger Spinner	early to late	evenings afternoons	warm days	Spinner	Diptera
22	Whirling Spinner	early to late	evenings afternoons	warm days	Spinner	Diptera

Chart Key to Enable Anglers to Fish Intelligently According to Time, Date and Rise

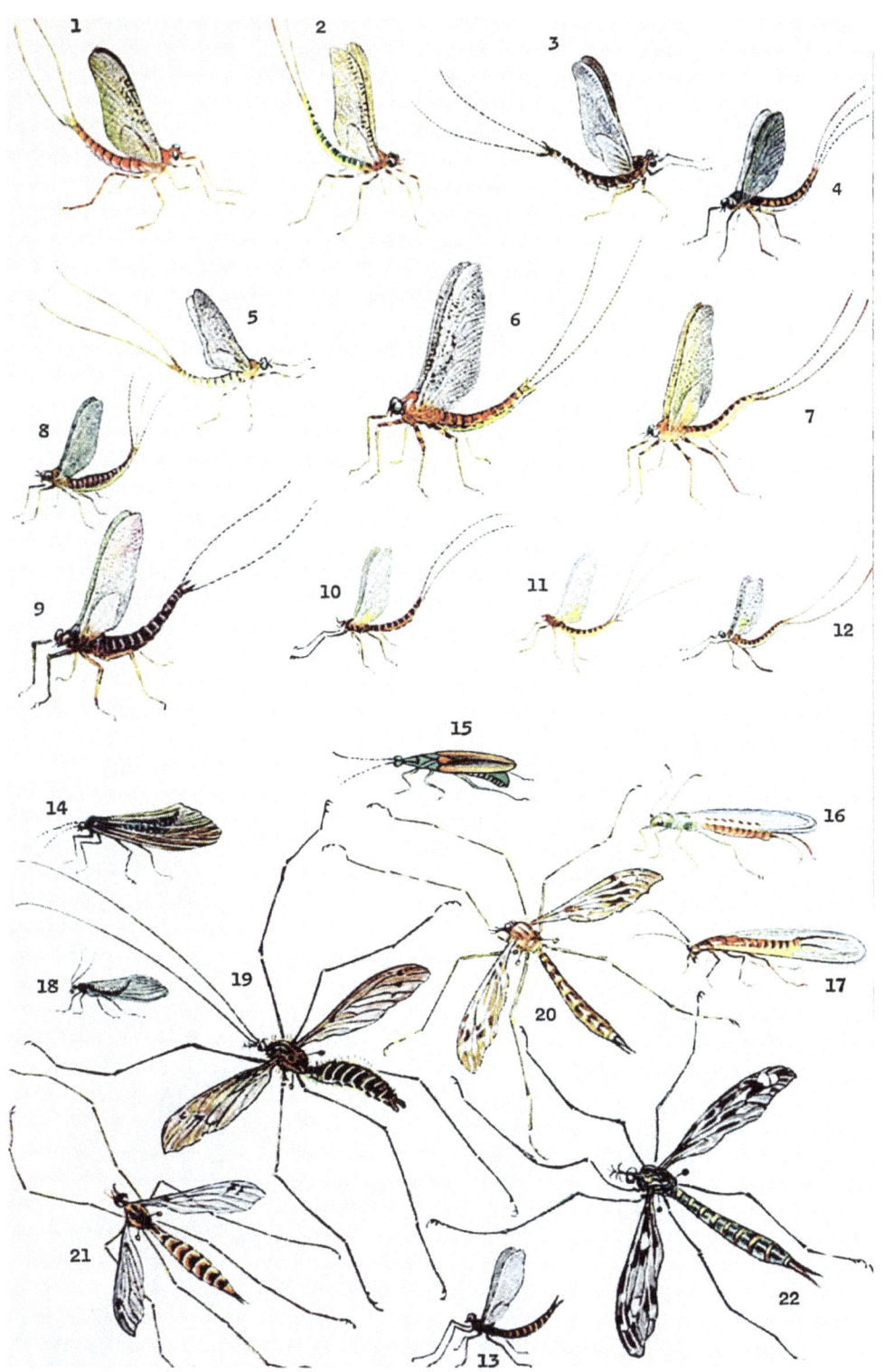

A SELECTION OF THE BEST TROUT INSECTS FOR THE MONTH OF JUNE PAINTED FROM LIFE BY THE AUTHOR

THE BEST TROUT INSECTS FOR JUNE

weeks of June, in 1913, 1914, and 1915, had three different weather conditions: hot, rain and wind, and frost and cold. It is these unfavorable conditions which I wish to overcome, as far as it is possible, by means of a better method in offering the flies and making them more true to the natural insect. Under normal conditions the pleasures of angling in June are greater than in May—though the basket may not be so full of fish. Nature is fully clothed, the mountain air is warm but crisp, and a thousand delights meet the eye at every turn —wild strawberries spatter the greensward with red, blossoms are everywhere, and tender are the young green leaves. The smaller trout ascend the brooks, and the big ones have the river to themselves, stately gliding from place to place, selecting for their lunch just what they choose. They get drowsy in the sultry afternoons; but at sunset, when in the fading light the river is alive with every kind of fly, they rise up, tempted by the feast, and gorge and gorge throughout the night.

Nos. 1 and 2. Female and male green-eye. These are the largest and most beautiful drakes of the month; in fact, of any month in the season. The female is heavier in the body and more highly colored than the male. Both sexes were caught the same evening, June 8th. In both, the wings are exactly like those of the green drake, a soft yellow-green. Both have large, brilliant green eyes, and both have shoulders of brownish pink. They are smaller in size than the May drake, but their actions in flight and at rest are quite similar.

The green-eye drakes are mostly evening flies, not overabundant, but a very tempting bait by reason of the repeated dips to the water, where they stay, floating beautifully like a graceful vessel with yellow sails, too long, alas! for their good. How I longed for an imitation of this lovely insect!

From sundown till dark I should use no other fly from the first to the twentieth of the month. Watch for a rising trout; fish upstream, if possible, and float this fly down a runway toward you. Then see the instant flash of the silvery sides of a big rainbow or the red spotted belly of a big native trout.

No. 3. Broadtail. So named from its thick tail, which tapers down from the same thickness as the body. A curious feature of this drake (and some others) is that the two fore legs are raised quite high from the ground, to be used as feelers. The two pairs of hind legs are yellow; while the fore legs are dark brown, like the body and tail, the latter having nine segment markings of a silvery gray. The entire under body is a dark yellowish-purple, and the wings a purplish-slate color.

This fine insect is an excellent day fly on cold or wet, windy days. It was captured from a fairly plentiful rise on a damp, cold day, and is therefore

THE BEST TROUT INSECTS FOR JUNE

most useful in such weather. It comes out quite late at evening, and may be used at that time, unless the green-eye is upon the water in goodly numbers.

No. 4. Greenback. So called because the upper body is green. It is a smaller fly than the broadtail; but, on the whole, it is more abundant on fair days and evenings. It can be used on warm evenings throughout the month. The legs are long, and the body stands quite high for so small an insect.

No. 5. Yellow-tip. An extremely dainty little insect, quite unusual in its markings. It has a dark yellow body, and a pale yellow tail, perfectly white underneath, the end tipped with bright yellow, from which shoot two long yellow wisks. The eyes shine out in vivid green, making it perhaps the most beautifully marked insect of this class. It is somewhat tender, and hides itself for protection from the cold; but it soon appears when the sun and warmth come back.

The artificial can be used at any time all through the month.

No. 6. Spot-wing. A large, handsome insect, quite plentiful on warm afternoons and evenings. The bluish, mottled wings stand high up on the rich, orange-colored shoulders. The entire body underneath is pale greenish-brown. The legs are yellow, mottled in brown-black, and the eyes are a vivid green.

The artificial can be used throughout the month when the weather is warm, and at evenings under any weather conditions.

No. 7. Lemon-tail. A slight, delicate insect, standing on high, mottled legs. This insect was caught early in the month and is over the water at all times, day and evening, and will surely prove a good small fly.

No. 8. Shiny-tail. So named from the shiny appearance of tail and body. The under part of the entire body is a dirty gray-yellow. The wings are quite dark in tone; indeed, the insect while in flight appears much darker than it really is. It is a cold weather fly, more abundant during the rain than at any other time.

One of the strange features in flies of this month, and, in lesser degree, those of the other months, is that light-winged and -bodied insects are more in evidence on the water in warm weather, and dark insects are usual in cold weather or rain.

No. 9. Chocolate. So named because the general tone of the insect is that color. It is a large and very fine insect, quite abundant late afternoons and evenings.

I should place the artificial of this insect first on the list of June flies for dark days, wet or cold, because of its size and plump appearance and its habit of constantly dipping to the surface after a few moments in the air. This insect, along with the green-

THE BEST TROUT INSECTS FOR JUNE

eye (No. 1), if only fairly imitated will entice trout day and evening through the entire month.

No. 10. Orange-black. A dark, natty little insect, seen all through the day rapidly flitting over the surface, now and then to dip, but always remaining quite near the surface. The remarkable contrast in color—very dark brown above and bright yellow beneath, fore legs dark and hind legs light—makes it a most distinctive object when seen on the water; and the importance of having the under body light is more evident in this than in any other insect I know that is at all plentiful.

No. 11. Tawny drake. A little insect entirely of one color, legs, body and tail being a dull orange, except that the under body is pale.

No. 12. Blackhead. A small fly with yellow under body, legs and wisks. The head is deep black, in striking contrast to the rest of the insect, which is two tones of yellow.

No. 13. Big-eye. This is another drake with some unusual features. The two fore legs are raised high toward the head, which seems to be all eyes. The body is dark brown and the wings are a dull slate color. The big-eye has no wisks, which makes it appear very bald and ungraceful in comparison with the other more beautiful species. It is fairly abundant all through the month; and I picture it only because of the peculiar characteristics so different from the rest.

With such a splendid array of the drake class to choose from (there were nine others captured) it was the most difficult thing to select those best calculated to kill trout. Nearly all were exceedingly plentiful at different times during the month. The fish, I am sure, were so fully gorged that any other creature but a fish would have been satisfied and wanted to sleep off the effects of fulness. I cut open a young brown trout, fourteen inches long, and found it to be stuffed full to above the gills with nothing but insects. It is nearly certain that trout in June refrain from fish diet, as the insects are much thicker over the water at night than in the day.

The two duns pictured are selected from seven specimens I caught in June; the rest being so similar that two are quite representative.

No. 14. Pointed-tail dun. This is the most abundant and the largest size of any of this class seen in June. They are all plentiful throughout the month, in all kinds of weather, both day and evening. It will be found a useful fly to alternate with the drakes; and it can be used as a floating dry fly, or by the wet method as second fly with either of the two browns as end fly.

No. 18. Black dun. One of a number of specimens very slightly different but all quite abundant at all times of day during the whole month. An

THE BEST TROUT INSECTS FOR JUNE

imitation will be found useful at times, as a change to a small dark fly.

No. 15. Goldrim. A pretty little land fly that frequents the water at evening. Its shoulders, tail, and eyes are green. The wings are a dull gray edged with golden yellow.

No. 16. Emerald. In form and mode of flight this is not unlike the yellow sally, but the varied bright colors are placed differently. The head and shoulders of this insect are bright emerald; the tail brilliant yellow; the upper part of the legs green, the lower part yellow. Staring and standing out of the small green head are two deep black eyes, from which shoot up a pair of horns in the shape of a lyre.

My imitation of this insect, with flat, white wings reaching far over the tail, was used with excellent results in July, evidently being taken by the fish for the light-colored browns so numerous in that month.

No. 17. Little yellow stone. This little fly is exceedingly bright in yellow and orange, being no doubt a newly hatched insect. It is somewhat larger than the emerald and the yellow sally, to both of which there is quite a resemblance when seen in flight. A larger specimen of this same insect was pictured in May, but it was not nearly so brilliant in color. Either one of these two insects should be included in a list of the "best flies" for most all conditions.

Of sixteen different species of spinners seen in June I select four of the most beautiful and abundant, and those which I think will prove killers with the patterns I have tied. Many species were swarming over the water on warm days and evenings, and I saw trout constantly taking them under as they alighted on the surface.

No. 19. Hairy spinner. This is quite similar in shape and size to one pictured in May; the body, wings, and horns of the two are alike, but in color this is more brown, and the head is less than half the size of the May spinner. The body and tail of this species are quite hairy, as is also the first section of the powerful legs. It is swift in flight, keeping quite low, near the surface, and moving round and round in circles.

No. 20. Goldbody spinner. With orange mottled wings, and the tail, legs, and shoulders a still brighter golden orange. The black eyes seem jammed down into the body, and the tip of the tail is black. This is a fine insect, but very hard to capture because of its rapid flight.

No. 21. Tiger spinner. This is a rapid flying and somewhat smaller insect. It has a bright yellow under body, greenish yellow undertail, and is all over a mixture of black and orange.

No. 22. Whirling spinner. An insect which

THE BEST TROUT INSECTS FOR JUNE

flies at astonishing speed over the rougher water, quite near the surface, ascending and descending, then whirling round. The wings are grayish, mottled beautifully in brown. The head and tail are greenish-brown, with a band of gold in each segment. The under body and tail are light gray.

If we make a comparison of June insects with May and July, we find that June has a preponderance of drakes and spinners, and May of two-wings, browns, and duns. In July we find a falling off in all classes of the larger and more desirable insects, worthy of or useful for imitations. This goes to show that the trout angler, to get the best results in fine sport and full bag of fish, must endeavor to be on the streams from the last week in April to the first week in July, at the latest, trusting to favorable weather conditions for August fishing—that is, just after copious rains, or cold spells.

The necessity for careful fishing and attention to the right kind of imitation is more important in June than in May or July; and for that reason a larger selection of flies from the June page is advisable. I selected eight flies for May, and I think at least ten for June. Four drakes, two spinners, two browns, and two duns will suffice for any weather or time of day.

It will often happen that some of the insects pictured in May continue to rise in large numbers during the first week, or longer, in June; and a few of the May flies should be reserved for that emergency.

The angler must not imagine that the great similarity of insects warrants the use of one for some other months. This difference is more apparent in the under bodies, the color of feet and the length and number of wisks. Thus it is that, while the upper body and color of wings should if possible be like the insect, it is of much less importance than that those parts of the insect visible to the trout be right.

In June the fisherman who is expert in casting the dry fly has the greatest advantage. Both the drakes and spinners should be played dry on the surface as much as possible.

Remember, the first or second cast will get a rise, if a rise is to be got in such a place; if no rise appears, the proper thing to do is to move to other localities. I do not believe in whipping one spot for half an hour in vain hope, unless a change of fly is tried. In other words, should the correct fly not result in getting trout to respond, nothing more can be done at such a place or time. Should you observe the green-eye on the wing at evening, try it for a few casts. If no response instantly occurs, put on a small drake. If the trout is still shy, try a spinner; then a dun. This procedure is meant

THE BEST TROUT INSECTS FOR JUNE

for those adverse conditions so often experienced in hot weather: scarcity of fish, overfishing, or too much natural food, with the trout gorged and off feeding on the surface flies.

VII

TYPICAL INSECTS OF JULY

As we proceed toward the end of the season the wisdom of dividing trout insects month by month becomes more and more apparent, for the reason that insects seem to have every month a distinct difference in each of the varied classes. The very same reason why trout refuse to rise while the hot sun pours its rays on the depleted waters also makes aquatic insects scarce in July. They are comparatively cold-blooded: they do not like the sun. I waded six miles on various hot days during the month and very seldom did I observe insects larger than midges till the sun was near setting. If a sudden change to gray, colder or rainy days came, a rise of flies suddenly appeared. This, of course, is the normal state of things; but July of 1915 was certainly abnormal, as were June, May and April, in weather conditions.

But these chapters are compiled partly from my notes and sketches running back some years to the time when I began the study of these insects. The observations of a single season would not suffice to cover the subject adequately in so erratic a climate as ours. Though, for that matter, nature equalizes things; and it cannot be doubted that these same insects have for centuries bred and lived on the same streams where we see them to-day, with varying abundance according to climatic conditions.

Old residents of the village of Roscoe, N. Y., tell me they could go to any part of the river (Beaverkill) forty years ago and fish in the most primitive way with worms, at any part of the season, and capture a large basket of fine native speckled trout in less than an hour. This is sure proof that natural insects were exceedingly abundant to keep the fish in good condition and so plentiful.

I can recall no July for many years with so much rainfall as that of the abnormal 1915 in the first three weeks of the month; for that reason the water was fairly cold and comparatively high and clear all the time. Sport was still unusually excellent in the daytime, both in the way trout rose to the lures offered and in the splendid gameness and ample size of the fish, even though well gorged with the fatness of June. I hooked more, and much larger-sized, trout than I ever remember.

Perhaps I may be considered impertinent to ascribe this to the use of my own tied flies. I shall have much better confidence after a thorough test; at least it will be more conclusive after using the right fly for the right month under more normal climatic conditions in future seasons. I fully realize it to be a bold assertion to infer, even, that everybody is using indifferent flies. Yet, again I repeat, if we copy nature, we must be right. The very reason why I undertake a most difficult work is to induce anglers to turn back to nature, to fact instead of fiction, to the living insect instead of the fancy fly. Had my own tied flies not been successful in rising trout up to my usual average I should be content to drop further effort and accept present conditions in the use of commercial flies now on the market.

On the other hand, I am more than ever convinced that in the last decade there has been a retrograde, or backward movement instead of forward, in the making of flies alluring to trout. In all the best one hundred artificial flies pictured in Mr. Halford's book, not a single one of them even faintly imitates any insect found on American streams. He frankly states that many of them are fancy flies, not intended to be copies from nature.

The most notable fact concerning the insects of July is the extraordinary abundance and variety of very small specimens of all three kinds, duns, drakes, and spinners. Sometimes the surface is alive with a moving mass of very small insects; then, in places, clouds of tiny mosquito-like insects are just as thick. They are, of course, no service

July Insect Chart

Numbers marked with asterisks are choice flies finely tied from the author's patterns and sold by his agents

No.	Name	Date of Rise	Time of Day	Weather	Family	Order
*1	Golden Drake	early to late	evenings and dull days	warm afternoons	Drake	Ephemera
*2	Pinktail Drake	middle to end	evenings and dull days	warm afternoons	Drake	Ephemera
3	Silver-Gray	early to late	evenings	warm evenings	Drake	Ephemera
*4	Spot-Tail	early to late	evenings	warm evenings	Drake	Ephemera
5	Little Orange Drake	early	evenings	warm days	Drake	Ephemera
*6	Olive Drake	early to late	afternoons	warm days	Drake	Ephemera
*7	Orange Stone	early	afternoons	warm days	Stone-fly	Perlidae
8	Brown Stone	early	evenings	warm days	Stone-fly	Perlidae
*9	Redhead Gnat	early to late	all day	dull days	Two-wing	Diptera
*10	White Miller	early	dull days and evenings	dull days	Moth	Lepidoptera
11	Tiger Beetle	early	all day	any time	Beetle	Colesptera
*12	Plume Spinner	early	all day	warm all day	Spinner	Diptera
*13	Golden Spinner	early	all day	warm all day	Spinner	Diptera
14	Green-Wing Done	late	evenings	wet days	Stone-fly	Perlidae
15	Orange Spinner	early to late	all day	wet days	Spinner	Diptera
16	Brown Bottle Fly	early to late	all day	warm days	Two-wing	Diptera
*17	Orange Miller	early	evenings and dull days	dull days	Moth	Lepidoptera

Chart Key to Enable Anglers to Fish Intelligently According to Time, Date and Rise

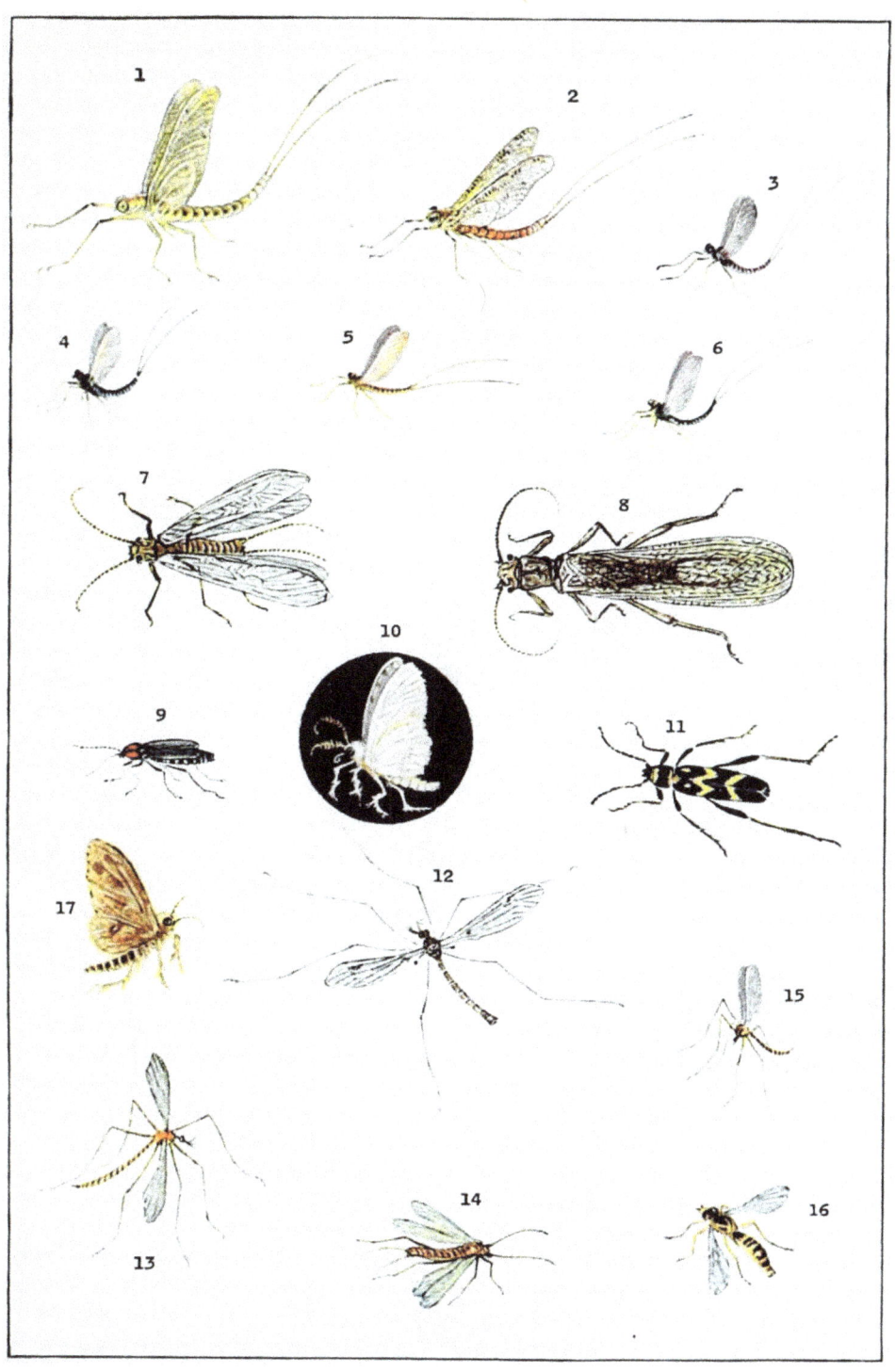

A SELECTION OF THE BEST TROUT INSECTS FOR THE MONTH OF JULY PAINTED FROM LIFE BY THE AUTHOR

to our purpose, yet no doubt they play an important part in the trout's menu.

Under the usual normal conditions in July, most flies are on the stream from just after sunrise, while the morning mists yet hide the sun, till about ten or eleven o'clock. Quit fishing then, and resume from six till dusk—by far the best sport of the day, because flies are most plentiful and trout visibly feed on them.

What I have named the red gnat is the most abundant day fly in any weather during July; and trout were feeding on it. Also many small flat-wing duns of a similar size were on the rise. So, too, were many tiny drakes and spinners, all in a mixed mass, flying over the water at every turn after sundown.

No. 1. Golden drake. This I would see sailing majestically along, a large, beautiful, solitary insect, flying sometimes low, with a dip, to again rise high over the water. Because of its similar actions the golden drake might be termed the May fly of July, though it is not half so abundant as the other species that appears in May. It is more plentiful after sundown; and from its very light lemon color I can distinguish it flying after dark, and so I assume that it continues in flight throughout the night.

Though very similar to the green-eye of June,

a careful inspection will show that the distribution of color is quite different on the upper part of the body; viewed from below the only difference between the male green-eye and the golden drake is that the former has mottled legs and vivid green eyes.

The artificial imitation can be used all through the month, day or evening; if day, it must be cloudy, without sun. A glance at the colored representation will show what a superbly made creature it is in form and color. Can any angler name a single fancy fly one-half so exquisitely lovely from any or every standpoint?

No. 2. Pinktail. The next important July insect is the pretty pinktail, with three sections of the tail tip a pale cream white, the rest a bright pink on top and under. The eyes are black, with spots of the same color on the yellow legs and the yellowish wings.

It must be noticed that these two flies, so similar to those of June, have no green eyes. The eyes are light in color, with black in the center.

No. 3. Silver-gray. This insect is very plentiful at sunset, but is not out in the daytime. It has large, round eyes, which stand out and up from the shoulders; and a tail hanging down, with the end bent up in a sudden curve.

No. 4. Spot-tail. Its eyes are large and brown; and its under body and its tail are colored a pale blue, spotted in black. This is another of the small-sized drakes seen only after sunset.

No. 5. Little orange drake. Like many other small specimens, it is very plentiful. Though quite small, its bright color and yellow under body make it conspicuous.

No. 6. Olive drake. The thorax of this little insect, and the top of its tail and wisks are a dark olive color. This also is an evening fly; and completes a round half-dozen selected very carefully from thirteen species, more or less alike in size, color, and shape, yet different in one or more features.

As before stated, these drakes are all evening flies except on dark or rainy days. This does not mean that the artificial must not be tried. I think it quite possible that, late in the season, it is an excellent plan to try some of the best evening flies in the daytime. I have succeeded on numerous occasions to rouse up a fish from its noonday rest by a tempting evening fly.

Nos. 7 and 8. Orange stone and brown stone. These are two stone-flies, slightly different from those seen in previous months; but the under bodies are so much alike as to render the little difference in shape not important enough to make an artificial copy. It will be noticed that the brown stone is considerably larger in size than others seen before. In August they are larger still.

No. 9. Redhead gnat. Captured early in the month, yet not seen in June, this is a good all-round fly for hot days, when few if any others are about. It is slow in flight, and the deep blackness of its body makes it very conspicuous on water or on land—it is seen in both places in the daytime. Except right under the mouth, which is red, the under body is black; the wings are slightly gray-black; and on top of the head, encircled in black, is a brilliant red patch.

For some days anglers on the stream were asking me for a black gnat or redtag, proof positive they observed trout feeding on this little black insect. I should place it in the forefront for July days, and I have taken much pains to make a good imitation. When in flight it is singularly like the wingless hackle gnat made in England. It is not a floater; so I think it will be more effective as a wet fly.

Though aquatic insects of July are so scarce, compared with May and June, the land is swarming with butterflies of every size and color. They cross and recross the rivers; grasshoppers in myriads skip on before you in passing through a meadow; numberless are the night moths that begin to wing their flight as the sun goes down; especially conspicuous are the white and yellow millers as they flick over the water time and time again.

No. 10. White miller. On dark days, and even

very rainy days, this insect is fairly plentiful in the daytime. Its body is quite fleshy, and though encumbered with thick, hairy legs of snowy whiteness, it often dips on the surface and seems to be able to recover and rise again in flight with apparent ease. Though I have never seen it float any distance, its attitude while on the surface for a short time, with wings close together, is quite similar to the drakes. Under a magnifying glass it is a marvelously beautiful insect, with golden fern-like horns, and big green eyes embedded in the whitest silken floss.

The white miller, as made, is absurd. No wonder one never gets a rise on a fly of that sort.

No. 11. Tiger beetle. This black and yellow beetle I caught in various sizes, with deep yellow and brown body and legs. The one pictured is deep blue-black all over, with markings in bright lemon yellow, except that the end sections of all legs are bright brown.

The big spinner, so plentiful in June, were all gone, and were replaced with many varieties of small ones. Some, indeed, were exactly like June spinners, only considerably smaller in size. They were mixing freely with the rise of small duns or drakes, or both.

No. 12. Plume spinner. This little spinner is very abundant, taking the place in trout diet of the same sized drakes and duns, now absent in the daytime. It is an excellent floater and rarely leaves the water's surface, where it skims around within an inch of the water. I never saw it ascend into the air. It has a pair of grayish-brown wings, and the body is beautifully marked in white, black, and brown.

No. 13. Golden spinner. A still smaller insect, which I call the small golden spinner, has a general orange tone all over, the thorax being brighter than the tail. The legs are part black, part orange. The black head hangs away from the body.

No. 14. Green-wing. A very small, orange-colored stone-fly, with legs and horns brown and the wings a yellow-green. This is the smallest stone-fly observed during the season, as the brown stone (No. 8) is the largest.

No. 15. Orange spinner. The very small orange spinner is a light, delicate insect that flies round in almost every place during the daytime, then in company with the very small drakes at evening. Its general tone is orange, with a gray tail, orange-brown above.

No. 16. Brownbottle fly. This small two-wing fly is swift in movement, but stays mostly at the side of the stream. I captured many in the net along with other insects that varied considerably in size, some larger, some smaller than the specimen shown. The wings are a bluish cast; the head and

TYPICAL INSECTS OF JULY

eyes are black, with a yellow patch in front; the thorax is brown; and the tail has bars of brown on very bright orange.

No. 17. Orange miller. A similar-shaped miller, tinted a pale ochre, which I have tied, is seen quite as often as the white one, and both should be very effective on dark days and nights.

The brownbottle is placed on the page more as an example of that species of two-wing flies that is most abundant, and it makes a pretty artificial that might possibly be effective in day fishing where drakes and browns fail—though I have my doubts.

While the page of July insects is not by any means so good a selection as those of May and June, there are at least six kinds that will surely be successful in luring trout in July: Nos. 1, 2, 7, 9, 10, and 12 are the very best flies of the month. No. 5, the small orange drake, should be held in reserve as a small fly.

It must be remembered that every one on the page for each month is good, and has been most carefully selected. One after the other has been withdrawn, till those left and pictured represent as nearly as possible a perfect selection for both day and evening, and any condition of weather that may happen from the beginning to the end of each month. Should it occur—and it often does—that

AMERICAN TROUT-STREAM INSECTS

an insect, pictured for the month previous, is in flight, it is because the rise of that particular insect continues over a month; indeed, some spring insects again have a rise in the fall.

VIII

SOME TROUT INSECTS FOR AUGUST

In most States of the temperate zone the trout season closes at the end of August. Some States have the close season at the end of July; in Connecticut it closes at the end of June. In all States the tendency is toward shortening instead of prolonging the season. This is a wise procedure, considering that the number of fishermen is more likely to grow than lessen.

I deem it best, therefore, to conclude with a list of flies for August, as it would serve no good purpose to include a page for September; though I believe that aquatic insects increase rather than diminish as the fall commences and hot weather abates.

In the temperate zone September finds most of the large female trout big with spawn, lying congregated at the mouths of brooks and springs of cold water, waiting for floods to carry them up to the breeding places. They are then in poor condition and unfortunately very easy to capture, and they are taken too often by unscrupulous natives, on worms or else by the baser method of driving them to shallow water and kicking or scooping them out on dry land without the aid of tackle or bait.

Many of the less important July insects continue to rise through the first half of August. Stoneflies were rising every day in large numbers, and they increased in size. The first new typical specimen of drake, which I name the black dose, appeared on the 15th. General conditions were normal—hot, sultry days, with frequent local thunder showers just after sunset, generally at precisely the time insects began to rise. Native anglers believe that lightning puts down trout from feeding. I experienced this difficulty on three occasions during August. Flies continued to rise after the rainstorm passed by, but the trout would not respond to my lures, though I fished till dark.

It was rare, indeed, to find any insects worth recording during the daytime. One conspicuous exception was a remarkable rise and heavy flight of the greentail ant, which I shall describe more fully in its proper place.

It would frequently happen that a fairly good rise of insects appeared on wet days, earlier in the season; but not so in August. It made no difference what the weather was, insects rarely appeared in flight till nearly dusk; sometimes it was pitch dark, so that I could not see to capture them.

Therefore, August is not a good month for in-

SOME TROUT INSECTS FOR AUGUST

sects and consequently it is poor for fly-fishing, even in the higher altitudes. Overgorged with an abundance of food, the water's temperature so warm in shallow places as to drive the fish to the bottom of deep pools, in addition the summer's constant whipping over the water by many anglers, what big trout are left have been hooked hundreds of times, perhaps have been played time and time again, only to get free.

It is no wonder that trout are then extremely wise and cautious. Very few trout did I land during August. They would rise to the fly once; but many a time that first rise was a miss, and no amount of coaxing with a change of flies would induce them to rise a second time. They would come up to examine the artificial, then go back to deep water. They knew the artificial and let it alone. The only time when they got hooked and I landed them was between dusk and dark, and that was the only time I saw them feeding on insects.

It is noticeable that drakes predominate in this month, as they do in July, though they are smaller in size—with the exception of the black dose. They are very similar in form and color to the six examples pictured for July. I caught no browns or stone-flies, but saw at evening isolated specimens of a large size flying high in the air over the water. I saw only one species of the dun class and that was quite plentiful. No large spinners appeared, though there were present all the time small ones a little larger than mosquitoes. The most plentiful insects in the daytime were fair-sized gnats and two-wing flies.

No. 1. Black dose. This began its rise on a rainy day, the 15th, and continued thick for several days. The half-submerged stones at the water's edge were fairly covered with the flat larvæ of this insect; some creeping up the stones, others just splitting open with the insect emerging from the case. The larva is a deep, shiny black, with two white marks over the head. The insect comes out with wings fully matured. I picked out a number by the wings while the larva skin was still attached to the tail of the fly. These insects fly slowly along, going up high over the water, but they are easily captured in taking the first flight. They do not appear on bright, hot days till quite late at evening; on wet days they are thick on the water.

No. 2. Small pinktail. Somewhat similar to the much larger specimen of that name shown in July. The wings of this insect, however, are brighter yellow, and underneath the body the first three and last three sections are pale lemon-yellow, the middle three sections being pink on top and below. There is a black spot in the middle of the eye. Small pinktails are quite abundant, especially at evenings, and sometimes come out late afternoons.

No. 3. Green-ribbed drake. A very pretty in-

August Insect Chart

Numbers marked with asterisks are choice flies finely tied from the author's patterns and sold by his agents

No.	Name	Date of Rise	Time of Day	Weather	Family	Order
*1	Black Dose Drake	middle to late	evenings	wet days	Drake	Ephemera
*2	Small Pink Tail	middle to late	evenings	warm afternoons	Drake	Ephemera
3	Green-Ribbed Drake	early to late	late afternoons	wet days	Drake	Ephemera
4	Brown Tip	early	evenings	wet days	Drake	Ephemera
5	Speckled Orange	late	evenings	warm evenings	Drake	Ephemera
6	Spotted Drake	early to late	all day	warm days	Drake	Ephemera
7	August Dun	late	afternoons	dull and warm	Dun	Trichoptera
*8	Sage-Green Ant	late	all day	warm	Ant	Hymenoptera
*9	Brown Buzz	late and middle	day only	warm	Four Wing Fly	Hymenoptera
10	Gray Buzz	late and middle	day only	warm	Four Wing Fly	Hymenoptera
11	Black Spinner	late and middle	day only	warm	Four Wing Fly	Hymenoptera
*12a	Bent Gnat	late and middle	day only	warm	Two Wing Fly	Hymenoptera
*12b	Brown Gnat	early to late	day and evening	hot days	Two Wing Fly	Diptera
13	Green Spinner	early to late	day and evening	hot days	Spinner	Diptera
14	Fluffy Spinner	middle to late	evenings	warm days	Spinner	Diptera
15	Gray Hair Caterpillar	early	all day	warm days	Caterpillar	Diptera

Inasmuch as August insects are few and of less importance and the trout season is closed in most of the states, a colored plate is unnecessary when color details are fully described in the following chapter.

The lower section contains artificial imitations tied by the author. Though fairly representing the insects in form and color, they are very poor indeed as compared with the beautiful flies of the four preceding months, tied by professional fly makers for commercial use.

Chart Key to Enable Anglers to Fish Intelligently According to Time, Date and Rise

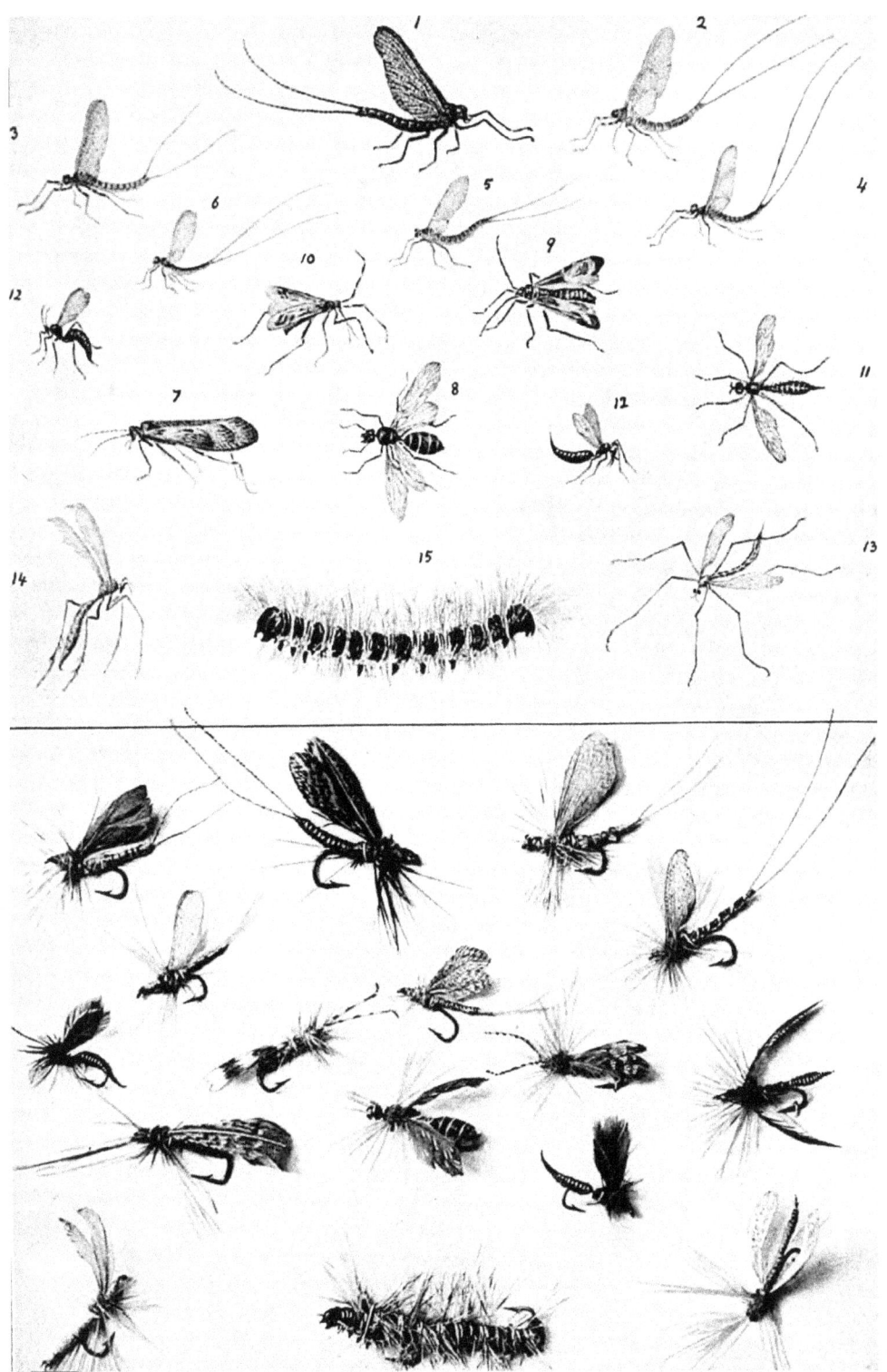

A SELECTION OF THE BEST TROUT INSECTS FOR THE MONTH OF AUGUST AND CORRESPONDING ARTIFICIAL FLIES TIED BY THE AUTHOR

SOME TROUT INSECTS FOR AUGUST

sect, with the tail sections ribbed in a greenish brown. This insect is out in late afternoons on dull or rainy days; on warm days it appears quite late, just before dark.

No. 4. Brown tip. A small, beautiful insect, with rich brown shoulders, and a patch of the same color at the top and end of the tail, the middle sections being ribbed on a greenish ground. The wisks are very long and are a pale yellow color, like the legs. This insect is fairly plentiful, and is quick in flight; though I captured many specimens in the net during the rain. A very unusual feature is that they vary considerably in size.

No. 5. Speckled orange drake. Another extremely beautiful insect, with metallic wings speckled in brown. This drake was caught as late as the 17th, on a cloudy day, and I saw it evenings only in warm weather.

No. 6. Little spotted drake. Though quite small its body is plump. It flies around all day in wavy motions over the water, dipping at the surface quite frequently. The general tone of the body and tail is orange, with bright yellow underneath.

No. 7. August dun. This insect, mottled in brown and blue, is the only dun of any fair size observed during this month. It is fairly plentiful in the afternoons and evenings during the latter part of the month.

No. 8. Sage-green ant. So called because the

general tone of the body and wings is a dull sage-green. Unlike the black ant of spring, this insect has a fat, shiny body of four sections; the shoulders and head being a little blacker in tone, and the legs a bright brown.

I consider this insect the most important of the month for the reason that on the 22nd day occurred an extraordinary rise: the surface of the water swarmed with them from four in the afternoon till dark, and during that time the water fairly bubbled with the varied fishes feeding on them. Trout, chub, dace, and bass rose to the surface everywhere in such numbers as I had not seen since the shad rise in May. For three days following, the rise continued in diminishing numbers, and the fish continued to feed on them. This rise was not confined to a limited area of a few hundred feet, for I heard that it extended over six miles.

This ant is a very pretty insect, both flying and at rest. In flight it is similar in appearance to the cowdung; not so swift, but just a steady round and round motion, most always over or near the water.

I captured many specimens which varied slightly in size. Some were attached to a smaller winged ant, which I imagine to be the opposite sex.

The shape and the color of this ant were so different from any artificial I had with me that I failed to get a rise to the flies offered. That is usually the case when very heavy flights occur. Except when

SOME TROUT INSECTS FOR AUGUST

the artificial is floated on the surface down a runway where the water is slightly rough, there is little or no chance.

Of twelve specimens caught in my net on the wing, ten escaped through the wire mess of the cage while I trudged along home with my prize—such a thing no other insect ever had sense enough to do. I went out again at dusk to capture a new supply, and these I covered over carefully till morning. When I got up I found all had cast their wings. Not to be outdone by these elusive little devils, I took sketching materials and pinned them fast while I made the sketches at the water's side.

This ant is a fairly good floater, and I shall imitate it with wings outstretched, also with wings lying flat over the body, both attitudes it assumes on the surface.

I consider myself very fortunate to have witnessed this rise, because it is, in a measure, a new discovery. I did not see a rise of the red ant, which is slightly smaller than the black ant; though neither of them by comparison is so good for imitation. I shall make an effort another season to be on the river at the right time of the rise, and I fancy there will be sport.

The remaining insects are of less importance; still, sufficiently important to make imitations, because they are selected as the best from many others less abundant.

AMERICAN TROUT-STREAM INSECTS

Nos. 9 and 10. These are, I fancy, male and female. They are quite common day flies in the latter part of the month, flitting about near the bushes, but always over the water. No. 9 is mottled in varicolored browns; the horns, divided into eight sections, and the very long hind legs, are all bright brown in color. They are swift in flight, which makes a kind of buzz, like a bluebottle. No. 10 is exactly like No. 9 in shape, but it is a little smaller and the color is varicolored greens and grays.

They are four-wing flies and are quite plentiful enough to warrant an imitation being made and tried.

No. 11. A deep black, shiny insect, covered with hairs. It seems to be of the spinner family. It has two wings of a dull gray; the ends of its legs are black, the inner sections a bright brown. It stays over the water and is very rapid in flight.

No. 12. A small, deep black gnat, very plentiful both day and evening. This is one of the rare insects to be out on hot days. I find at this late season when trout will not be persuaded to come up after large flies, these little gnats often succeed in enticing them to rise. There is another gnat of a similar shape, but brown in general tone. It is a pretty fly, likely to be successful, as it stays over the water and is very rapid in flight near the surface.

SOME TROUT INSECTS FOR AUGUST

No. 13. This is the only small spinner I saw in flight that was at all plentiful. Some large ones, isolated specimens, were the same as those pictured in July.

No. 14. Fluffy spinners. A two-winged, moth-like insect, of a pale creamy color, with very long legs and tail. Its entire body and wings are downy. One would take it for a spinner were it not that its flight is quite slow.

No. 15. This is an extremely abundant hairy caterpillar that is so thick in numbers as to be crushed at almost every step along the edge of the stream. I introduce it this month because I think a good imitation caterpillar would be very effective for big trout—possibly more so than flies in August. In fact, every month has some particular abundant caterpillar. There is another green hairy caterpillar in June, quite as plentiful, which breeds in a sort of webbed enclosure on the apple-trees. During the floods in May there is the common brown and also the black hairy caterpillar. These three I did not include in those months because the space was needed for the overabundance of insects; and I am doubtful if they would at that time prove such effective lures (except in floods) as this August specimen would be when insects are comparatively scarce.

Before concluding this list of monthly insects I would ask the prudent angler to give a fair test to a selection of the artificials. By a fair test—I care not what water it be, if within the temperate zone—I mean that the angler can choose any one single fly and cast it where trout lie, at the right time mentioned in the fly chart, and my firm belief is that, if played as directed, it will attract and hook more trout, as well as give more universal satisfaction, than any six varieties of fancy flies cast in a "chuck and chance it" style. It must be so, unless trout get tired of their own natural food to prefer something different, which seems to me improbable. My sole aim is cunningly to deceive the trout: that is, to induce the fish to imagine it is really taking its natural food, in the form of an artificial true in color, shape, and size.

Concerning tests being made of these new flies in more northern waters—that is, in Maine, New Brunswick and western Canada, including also the Northern Pacific States—I shall look forward to such tests with deep interest. I shall make the tests myself at the earliest date, at the same time making a collection of sketches of the insects native to those regions. The difference of temperature will, I think, have little effect on aquatic insects of the two zones, especially about midsummer. The difference will be greater early and late in the season.

IX

SIX BEST FLIES FOR EACH MONTH

In order that anglers may not be overburdened with too many flies for each month, I choose from each monthly plate a carefully selected list of about six of the very best, that are fairly representative and will give the most successful results.

While the five monthly plates represent nearly one hundred varieties that were carefully chosen from probably over five hundred different specimens, I am of the opinion that this restricted list should be known to the angler, that he may provide duplicates and avoid confusion of many flies.

Some anglers will consider that six flies are a very limited selection to take along on a fishing trip. They would be, were it not that each one is sure to be effective, and that one fly may be successful all day long—and perhaps for many days.

AMERICAN TROUT-STREAM INSECTS

This selected list is intended for those who fish with a single fly—dry or wet. I used to fish with two, sometimes three, on a cast. After trying all kinds of dodges to deceive trout, I go back to the single fly, casting carefully at the right time in the right place, with frequent changes, if required. When trout are repeatedly rising to a heavy flight of a certain insect upon the surface, one would imagine that three imitations, exactly alike, would be most likely to capture a double, or at least be more effective than a single imitation. But the latter invariably proves the best rule. If three small wet flies attached to a long leader be allowed to dash sunk along a swift runway, one of them will often be taken, usually the tail or end fly; the other two act as teasers. For really fine fishing, looped snell flies—however far apart they may be attached to the leader—destroy a good accurate cast, instead of assisting, as one fly does. If trout are feeding, a single fly, well cast, within the vision of the trout, will be more likely to entice it to rise than any number of flies on a single cast.

APRIL

The flies I recommend for this month's selected list are Nos. 1, 4, 8, 9, 10, and 17.

No. 1 is a little black fly, most useful at all times, wet or dry, mornings or evenings. If the water is

SIX BEST FLIES FOR EACH MONTH

high and the weather warm, it would prove effective as second fly to the brown drake, No. 4.

This No. 4 appears late in April, but can be used early if the days are warm and sunny. You can always rely on the brown drake, whatever the conditions are in this month, as trout will not refuse it if you place it anywhere within their vision.

No. 8 has been tested and found excellent on bright days.

No. 9 is very good on dull, wet days, any time in April or May. It should be fished at the surface; though I have succeeded in getting fish to rise in almost every condition.

No. 10 is best on warm, windy days, on rough water; also at the foot of pools, either floated or sunk. This fly is good as an end fly, with No. 1 as upper fly, both sunk in the deeper parts of the stream. I invariably make a practice of keeping sunk flies on the jump—never still or in one place —with frequent casts all over the water.

No. 17 is the female shad-fly without eggs, and it can be used any time under any condition—wet or dry, warm or cold; but it must be used always at the surface.

The angler should provide himself with at least three of each of these six flies. Even an expert is liable to lose or destroy his flies; therefore duplicates in reserve come in handy.

MAY

For this month I should extend the list to eight best flies: the variety is greater and the fish will rise more readily, as it is the best month for trout fishing with flies. You need plenty of duplicates; for the weather is most always erratic, making it necessary to change flies quite often.

The best eight flies are Nos. 1, 2, 6, 8, 10, 12, 17, and the male and female shad-fly.

On a warm day in May you can fish with No. 1, green drake, all day and evening; though it is advisable to change off now and then with No. 2, brown drake, if you are not getting the fish to respond to the former. Both should be floated all the time.

No. 6 is a dark, rainy day fly. It also is to be floated; and it is best mornings and evenings.

No. 8, the black ant, is good all through the month in any weather.

No. 10, yellow sally, though a warm day fly, can be fished all day in any weather; but it is best at evenings in warm weather. It should be floated; and fished in smooth water, either on level stretches or pools.

No. 12 is a good evening fly in any kind of weather.

No. 17 is the golden spinner and may be tried any time in sunshine; also it is very good at eve-

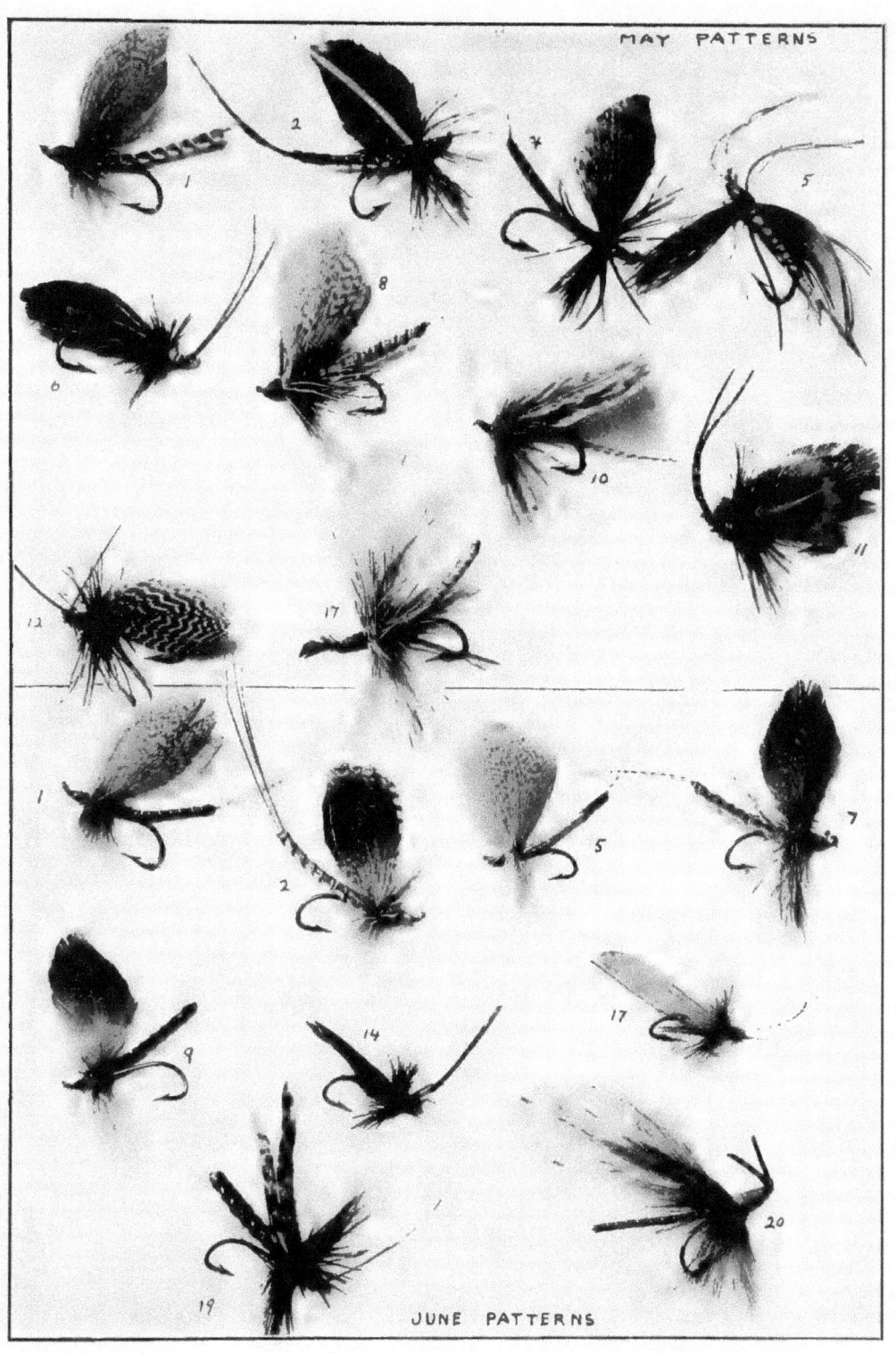

SELECTED NATURE FLIES TIED IN ACCORDANCE WITH THE AUTHOR'S
PATTERNS AND SOLD BY HIS AGENTS

SIX BEST FLIES FOR EACH MONTH

ning. It should be played at the surface in all places where trout are likely to be located.

The male and female (with egg-sack) shad-fly can be fished at the surface under any conditions throughout the entire month. When the rise appears thick, attach both male and female to the leader and cast them to the rising fish. The shad-fly is far superior to any other May fly.

JUNE

For the month of June I find a selection of ten will hardly suffice to do justice to so many beautiful examples. Out of forty-seven different species of drakes captured, I pictured thirteen; and now to sift them down to six is a difficult thing to do when each one has some unique feature of value peculiar to itself. June is the drake and spinner month; duns and other flat-wing flies suitable for imitation are comparatively rare.

For the purpose of variety I select Nos. 1, 2, 4, 5, 9, and 11 of the drakes; and Nos. 14, 16, 19, and 20 to represent the best of the other species.

The male and female green-eye furnish an excellent lure for warm evenings and afternoons of the first three weeks of the month. They should be fished floating on rough water, or swiftly but lightly moved over placid water.

No. 4 is good at all times; best at evenings just before dark.

No. 5 is a splendid all-round fly for any condition or time. But the angler must use judgment: if he discovers this fly on the wing, use it; if not, imitate that insect which is most abundant.

No. 9 is every bit as good as No. 1, and it is reliable at all times of day or evening because it is present during the entire month in fairly good numbers.

No. 11 is the same—a fine small fly that is taken at almost any time.

No. 14 is a dark dun, very useful for a change from the drakes, if the former are not out, or are not taken by the fish on cold, dull days.

No. 16 can be used all through the month; but, like many others, it is more effective from four o'clock till dark. Weather conditions play a large part in the choice of flies during June.

No. 19 is a large spinner, to be used during the afternoons and evenings. It is best cast on the surface in quick flights, but never left in one place and never allowed to sink.

No. 20 can be fished all day. It should be dragged along the surface and may be sunk in swift runways or guided over and round submerged rocks near rough water.

For successful results in June fishing a great deal depends upon favorable conditions. The best time is when there is plenty of water, and the day is warm and without wind. In every case, it is desirable

SIX BEST FLIES FOR EACH MONTH

that you fish carefully, casting upstream as much as possible.

JULY

In July it is still more difficult to get trout rising in the daytime, because so few flies are on the wing in the hot sunshine. There are short periods of colder, wet weather which brings out additional flies, and it is best to take advantage of their appearance. Early morning fishing before the sun is up, and evenings till dark gives the best sport in July. For that reason I shall choose but six insects. Nos. 1, 2, 7, 9, 10, and 12.

The most beautiful insect of the entire season is the golden drake, No. 1. It should be cast very lightly and floated in all likely spots where trout are known to hide. It is not overabundant, but it appears all through the month, at evening, and it is out also on dull or wet days.

No. 2 does not appear till the latter half of the month. It comes on the water at evening and on dull or wet days. It might be an effective fly early in the month also. Its very attractive color and its medium size make it a most valuable fly for July.

No. 7 is a medium-sized stone-fly. It is always useful as a change, and may be utilized almost any time during the month, day or evening.

No. 9 is a valuable small gnat for any and all occasions, making the best day fly of the month, for warm days especially. It can be played at the surface or sunk.

No. 10 is an evening or wet day fly, and should be floated or fluttered along the surface.

No. 12 is a small, though very abundant, spinner; and it is useful any time, day or evening, either floated or sunk.

AUGUST

For August, the last and least productive month of the season, I select six flies, nearly all small in size and very quiet in tone of color: Nos. 1, 2, 8, 9, 12, and 15.

No. 1, the black dose, is a wet day fly, very plentiful during the latter half of the month. On wet, gloomy days it rises and flies slowly along in goodly numbers; and on warm evenings it is a most excellent fly.

No. 2 is a beautiful, though small, pink drake. It is best at evening, but it is often out when the day is not too warm. It should be tried in the early morning before the sun is up.

No. 8 is the sage-green ant, probably the best fly of the entire month, though much depends upon the period of its rise. When it does appear for a period of two to five days all other flies can be laid aside. The sage-green ant may be floated or sunk.

No. 9 is a small daytime fly, sometimes very good for warm days.

SIX BEST FLIES FOR EACH MONTH

No. 12 is a little black gnat, useful for hot days and evenings.

No. 15 is a hairy caterpillar, very useful for deep pools and swift runways, sunk anywhere in deep water where trout hide from the sun.

Fishing in August depends entirely on the weather. If the water is extremely low, on hot sultry days the chances are very slim that you will get a rise to any fly whatever. At evening, conditions are vastly improved. Big trout begin to move to the shallows from deep water, after both minnows and flies, and that is the only time you can entice trout. After a good day's heavy rain—or, better still, a rain that lasts all night—the following morning, if cloudy, would be the most favorable condition for August fishing.

This chapter may possibly be considered somewhat of a repetition of the monthly list of flies, but the value of this selection is that it makes the list far more simplified and easier to follow.

I can well conceive that no angler will bother to carry along a book to consult while fishing, or even take it along on a trip, to find out the proper use of the various flies. But he can devise a plan, as I shall do: that is, to keep flies separate in different little boxes, marked *warm day flies, cold day flies, evening flies*, etc., etc. By dividing them up in this manner, we soon get familiar with the right time to use them; and of course the natural fly on the wing will be a never-failing guide. The descriptions are so accurate and the drawings are so faithfully copied that there will be no danger of not recognizing the natural insect on the wing.

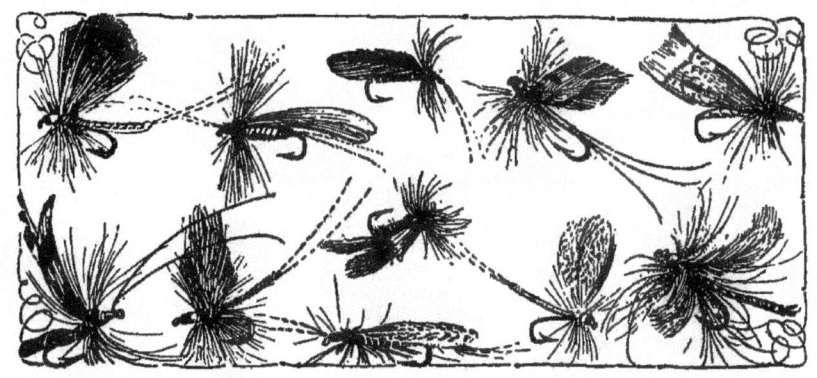

X

CONCERNING MY ARTIFICIAL IMITATIONS

Some time ago, when the material contained in this book was running in serial form through the pages of *Field and Stream,* I experienced the greatest difficulty in trying to persuade both amateur and professional fly-makers to tie patterns from my drawings of the natural insect. The excuses were both amusing and annoying. The only thing left for me to overcome such a setback to my work was a determination to tie my own patterns. The knowledge I had of the art of fly-tying was practically nil; but various friends who tie their own flies (and very good ones, too) seemed to think the hand and brain of an artist were sufficiently delicate and capable of doing so, and I set about the task to accomplish the object of my desire.

I have since had reason to be very grateful that those refusals led me into what I now consider the most delightful branch of our well-beloved recreation. True, it forced me into some months of close study during the hot weather; but the final outcome was a fascinating and agreeable surprise, and I am well repaid. I still have much to learn; but that's the fun of it. I am not envious of the fly-makers' superior workmanship, because I know that practice makes perfection if you devote sufficient time to the work. While I desire very much that anglers will get my artificials and test them, I also wish them to follow my footsteps and enjoy the pleasure of tying their own flies, either after my own drawing of the insect or from the natural insect itself. That is the reason I include a chapter in this book on how to tie artificial flies. What success I have achieved, aside from the work involved, is due to a great extent to Mr. Halford's excellent chapters on how to tie flies in "Trout-Stream Entomology"—a book every thoughtful American angler should read, if not possess; for it goes much farther than I do and is more elaborate in treatment.

When the patterns of nearly one hundred newly named flies were finally made, another serious trouble appeared as to how they could be supplied to anglers at a reasonable cost. The great number of communications from magazine readers and other anglers soon made it evident that I could not possibly alone supply enough samples of my own handi-

ARTIFICIAL IMITATIONS

work, and I set about securing a reliable firm to make the flies and sell them. By confining the manufacture of these new flies to one house, I could be more certain of a uniformity and correctness in copying the patterns, especially if I gave my personal supervision to the work. With that end in view, I arranged with the firm of William Mills & Son—agents for the famous Leonard Rods—Park Place, New York City, to have exclusive rights to make and sell all my new patterns which are ten for April, ten for May, ten for June, and ten for July. In that way only could I guarantee that the patterns are what I intend they should be.

It is very possible that copies will be made and sold by others; but only those flies got direct through Mills & Son will have my endorsement as correct copies of my drawings of the natural insects.

One of the dealers said to me, "We have many flies *nearly* like yours."

"That may be so," I replied; "but have your flies under bodies paler than the upper? Are they used when the natural insect (from which they are copied) is on the water?"

By the systematized method of fishing, success is sure. The haphazard method of casting any sort of fly without consideration of the feeding trout will, nine times out of ten, utterly fail; if you are fishing in anything but extraordinarily favorable

AMERICAN TROUT-STREAM INSECTS

conditions—viz., when trout are exceedingly voracious and food is scarce. The prevailing conditions are, alas, mostly the other way. That is why these new flies and new methods are more to be desired by the rank and file of anglers.

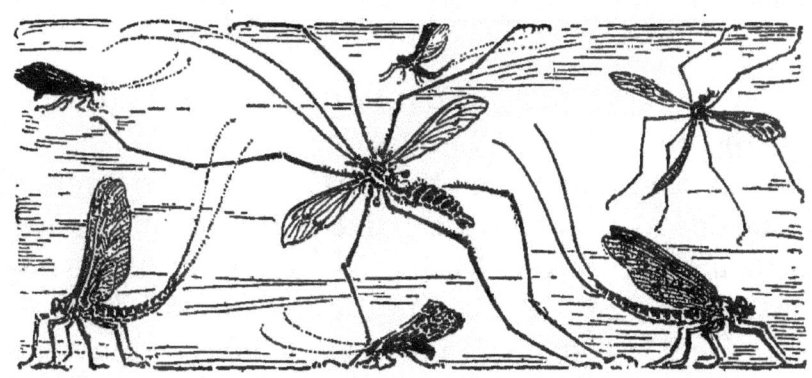

XI

NEW NAMES FOR FLIES

AFTER tying these artificial imitations from my drawings of the living insects, I found it necessary that a new name be given them in order to identify one from the other, both for my own use in future and to assist others who may perhaps want artificials copied true to the natural insect. This perplexing problem I solved in having the names chosen to denote some particular form or color of each individual insect in the various classes. To name flies after rivers, places, or people is provincial, commonplace, and utterly in bad taste. We could not follow the names found in English books on angling flies; nor the flies that by age are so well known as typical American flies. Assuming the insects pictured in English books on entomology or angling to be exact copies of nature, after diligent search I find no duplicate in form and color of the insects native to American streams.

For that and some other reasons it is far preferable to go along original lines, at least in the beginning, if we wish to simplify this important subject. To copy the natural insect faithfully, then to give it a common (and what I hope will become a familiar) name that is distinctly and typically American, will be a start in the right direction to attain order and system in classification.

It is out of my province as an artist angler to search out from books on entomology the Latin names of each species caught and pictured. The work is quite difficult enough as it is; and I doubt if it would serve any good purpose or that such names are desired by the humble followers of the immortal Izaak Walton, who, like myself, prefer to devote precious time to more useful things. I find upon careful inquiry that no book has yet been issued on the entomology of American aquatic insects alluring to trout, nor have these insects been even classified. I was asked by an angling expert who was examining my drawings, "Why don't you give the proper Latin names to each fly?" My answer was, "I would do so, but no entomologist has yet made any effort to classify American trout insects into orders or divisions, families and species, as has been done in France and England." It would take a lifetime of not one man, but several,

NEW NAMES FOR FLIES

to collect and classify a complete collection of the various orders of the insect world in so vast a country as ours. This wide field is waiting for some able student or professor of entomology.

The present conditions are now that the amateur—indeed, the average angler—is helpless: he must of necessity purchase whatever the dealer has for sale, both domestic and imported, in the way of trout flies; and it is natural to suppose that the dealer wants to buy low and sell at high prices. It is a remarkable fact that certain popular flies—like the March brown, cowdung, and many others I could mention—as sold by the dealers are as unlike the natural insect as possible. Each maker has his own idea what a March brown is; but, curiously enough, no imitation is at all like the natural insect of that name. So I change mine to the old-fashioned title of brown drake.

XII

THE MAKING OF AN ARTIFICIAL FLY

The manipulation of a trout fly is a much more simple undertaking than I had supposed. The circumstances which, in a measure, forced me into this most delightful art are noted in the chapter on "Artificial Imitations." Not everybody is constituted to make a fly; though the attributes of all "born anglers" conduce toward the successful accomplishment of this fascinating work. These attributes are patience, delicacy of touch, and a certain love for everything intimately connected with our favorite recreation.

Only the most salient points, condensed and briefly told, will be required for the beginner to get an insight of the art; the method is best learned by practice and experience.

There are no laws or rules governing fly-tying; the all-important thing is to get neatness, solidity,

THE MAKING OF AN ARTIFICIAL FLY

strength, and artistic finish by the use of the tying silk and in the winding up or finishing knot. Considerable deftness in cutting a neat, well-shaped wing, and a certain delicacy of touch in fashioning various shapes and forms of the body are required; and those who happen to have thin, pliable finger-ends have quite some advantage. The reason girls are best fitted to make flies is that their fingers are more apt to be long and thin.

The amateur fly-dresser should first read and study some of the excellent books devoted to the subject, the majority of which have been published in England. I learned most from Halford's "Dry Fly Entomology," which gives the different methods of a number of experts. I think that the beginner will find in this chapter sufficient information to make a start, with the aid of the diagram sketches, and then he should work out for himself a method most suited to his own individual need.

MATERIALS: FEATHERS AND HACKLES

The fly-dresser's first and most important work is to gather and get together a collection of feathers, particularly wing and hackle feathers, the latter being taken from the necks of different breeds of poultry and game birds. Hackles are the stiff feathers employed to imitate the legs of various trout insects.

There are two ways to gather a collection of

feathers, viz.: to buy them, and to beg them. The former is easier and quicker; but the latter, though slow, is far more interesting, because it is a hunt or quest, which is always interesting in many ways. You can "buttonhole" every poultry-man, hunter, furrier, and taxidermist of your acquaintance for assistance; and in that way you are more likely to get choice varieties and colors not available in the market at any price.

Get the hackles on the skin, if you can; for they are then bunched together and are more handy to use, and you can keep them better in stock. If you cannot get them on the skin, tie each variety in bunches, and keep them in glass jars, where they are quite secure from moths. If you are a smoker, you can use the pound glass tobacco jars (with covers), which are just the thing for the purpose.

Mr. Halford advises—and many fly-makers follow this method—that you trim off all the hackle and wing feathers, by stripping off the downy portions. This is for a dual purpose: first, moths go for the downy portion in preference to the feather; second, your feather and hackle are always ready for use to select from.

Personally, I do not agree. I have found out, after tying a hundred new patterns, that the lower parts of hackles are most useful for the long legs of spinners, and the lower parts of wing feathers

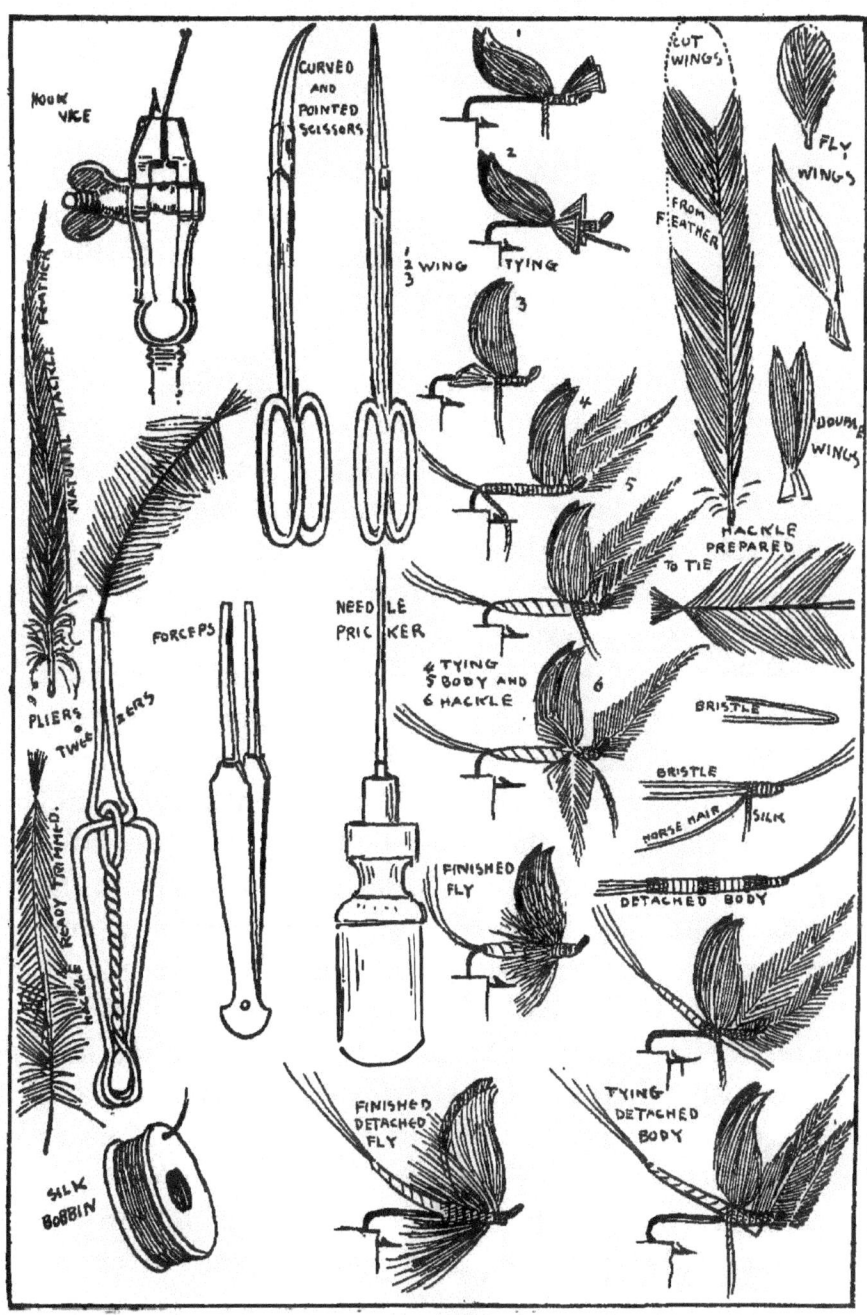

TOOLS FOR, AND METHODS OF MAKING A FLY

come in handy for many important uses—the quills being useful for detached bodies, and much of the downy portions for packing bodies, hairy insects or caterpillars, etc. Then, again, you never know in advance how long you may need a hackle or feather. For these reasons, I prefer to trim feathers at the time I want to use them.

Some flat stripped wing feathers are best kept in envelopes; but I much prefer to keep the wing intact, stripping as I go along. There are many small inside wing feathers which are most useful; in fact, every feather on the bird comes in at one time or another. Curved feathers should be kept in boxes or glass jars.

When you start your collection, remember that small birds are most useful in making little flies. The wings, tail, and neck feather of the cock sparrow are extremely useful; and the entire skin and wings of the starling are better still. Beautiful hackle feathers for small flies can be taken from the neck, rump, and under wing of the starling and other birds of a similar size. It is quite a simple matter to trap small birds and skin them immediately before the body becomes cold. When the skin is ripped off, just tack it (with feathers down) to a piece of board, and let it lie in a dark closet away from flies for a few days, and it will dry without any powder or solution. Large birds with fat on the skins can be sprinkled over with saltpeter after the

THE MAKING OF AN ARTIFICIAL FLY

fat portions which adhere to the skin are carefully shaved off.

One of the most useful parts of your feather stock are the fibers from the tail feathers of a peacock. This is called harl, and is useful in heads and bodies such as those of the fancy fly called coachman or black gnat. The harl is wound in precisely the same manner as silk or tinsel. In procuring harl fibers, be sure to get them newly plucked—if possible, from a living bird; because then they are strong and are bright in color, and they keep so for some time. Old peacock feathers turn from the vivid iridescent green color to a dirty brown.

If you desire to have some feathers dyed various colors, you can purchase some of the reliable packets of dye, following the directions given with the packet. This probably will prove the simplest way. I have not yet made any effort in this direction. I would rather trust to the natural-colored feathers; it may take a little more time in selection, but in dyeing feathers there will be some troubles to contend with, it is certain.

The beginner will know best, and have his own way to arrange to keep feathers from moths. It is certainly a most important thing to keep them clear; for when moths do get among them they destroy in a short time your choicest and best specimens. A good, tight, well-made box, or a small closet with shelves made especially to place all the materials for fly-tying, is the best thing to have. Any receptacle in which you place feathers should be opened and examined frequently, when not in use, so that ravages by these little pests may be curtailed and destroyed. In addition, moth powder should be sprinkled over everything in the box so often that the strong odor is ever present.

VARIOUS OTHER MATERIALS

The large department store's embroidery counter is a mine of wealth for the fly-tier. There may be found nearly everything you want in the way of plain and floss silks, gold and silver wire, raffia grass, and wool.

To begin with, you need a big spool (1500 yards) of black silk for winding. It must be a combination of fineness with strength. No cheap silk will suffice for the purpose. As I am satisfied with the silk I use, I give the name. It is Holland Company's Prime Black, No. 00. Upon this winding silk depends a good deal the fly's being strong and neat.

You will need a varied selection of smaller spools of colored silks; in fact, nearly every shade of color is desirable. These may be got in fifty-yard spools. They should be twisted silks of various thicknesses, and also selected colors of floss silks.

For the body windings, wool is of greater service than silk, as it retains its bright colors when wet.

THE MAKING OF AN ARTIFICIAL FLY

What is known as Lion Brand Saxony wool is a useful article, because the hanks are dyed so that each shade of green—or other color—fades into a darker or a lighter shade at about every yard-length; thus it is possible to cut off pieces any shade you want.

For foundation work I use mercerized cotton of different shades and thickness.

Another very important material is some mohair wool dyed in different bright colors, so necessary for the rough, hairy appearance required in certain flies.

Raffia comes dyed in almost every possible shade; and a sorted selection for ten cents will last a lifetime. This raffia is very useful indeed, for every part of the body, and it keeps it bright color in the water perfectly. It can be used flat over a foundation, it can be twisted for winding upon, or it can be used by itself as a body; and it is as strong as silk.

The gold and silver wire may be bought in many ways: flat, or round, or twisted, or flat tinsel. A good selection of all kinds and thicknesses is advisable; and all of them should be kept together in a box, from the light and dampness, so that time will not fade them.

All these various materials may be purchased together either at a department store or at a regular embroidery establishment.

For the legs and bodies of some flies, the hair or fur of certain animals is used. It can be bought for a few cents from the furrier, because the smallest clippings will suffice. White, black, brown, and gray are most needful; and an extended selection is best. It should be kept in a jar, protected from the light and made secure from moths.

In making detached bodies, or cocked tails of drakes, boars' bristles are used. These can be procured from shoemakers' supply stores; where you also may get the black shoemaker's wax used for waxing the black tying silk.

Another useful body material is the quill stripped from large feathers of peacock, eagle, and condor. This quill is carefully stripped off, and then cut thin or wide according to what is required. The quill can be dyed almost any color, and it is extremely useful in many ways. When tied round the bare hook or on a solid body foundation, it makes a splendid winding for some kinds of bodies.

White, black, and red horse hair, also, is used with excellent results for small, thin bodies or for the section markings on the tail. White horsehair can be dyed various colors and utilized with telling effect, especially in making the small tails of drakes.

The dressing of delicate-colored bodies requires a transparent wax that will not discolor the silk or the feathers that are wound.

For tail wisks, mottled or plain-colored fibers

THE MAKING OF AN ARTIFICIAL FLY

from the various wing and other feathers should be included.

Lastly, a small bottle of varnish to harden the tying-off. This will about complete the list of materials for the amateur or beginner.

IMPLEMENTS

The implements and tools required are few. A pair of small, sharp-pointed scissors; a pair of nail scissors with curved points; a thick, blunted needle, fixed in a wooden handle—mostly used for picking out the tying silk when making the half-hitch and making fast. A pair of hackle pliers or tweezers; a pair of smooth pointed forceps to pick up small objects or feathers; and a small vise to hold the hook.

A vise may be purchased from a watchmakers' tool supply shop. I bought one; but I soon found my left forefinger and thumb to be more satisfactory in every way. Halford considers a vise absolutely necessary, but he does say, "Many well-known amateur fly-makers prefer to use their fingers and work without a vise." I have also dispensed with hackle pliers; I can grasp and twist the feather around the body with far greater freedom and neatness. And I see very little use for the forceps, except to those who have large fingers. Mine are small, and I keep the nails of my two thumbs and first fingers longer than those of the other fingers, so that I can readily pick up any object with ease. All anglers will find it most convenient to keep those nails longer to facilitate tying knots, untying them on gut, or fastening eyed hooks to the cast.

Another needless implement is the small bobbin to weigh down the tying silk while you do the winding. I found the bobbin to interfere with both hook and feathers. If you wax the silk thoroughly, it will hold stiff and keep in its place.

The only tools I use are the scissors and the thick needle, the latter being constantly in service. I make mine out of a lady's hatpin, cutting it to about three inches long, and fastening it in a small wooden handle.

To tie flies for any length of time in one attitude becomes wearisome. To make it easier, I have all the necessary implements and materials neatly arranged in a box, of either cardboard or light wood, two inches deep, and twelve by fourteen wide. By this means I can move from place to place or at least take a different position and go on working without getting tired. If you get used to a vise fastened to a table, on to any fixture, you are bound to it—and to a backache.

MANIPULATION

The amateur must first understand that in the making of artificial flies the method of procedure

THE MAKING OF AN ARTIFICIAL FLY

varies considerably with each individual. No two persons tie alike; no two seem to dress a certain fly the same, and I venture to assume that of a number of fly-dressers copying the natural insect, nearly all would go about it in a different way, and would have an entirely different result. At the age of five—that is fifty years ago—I began under my father's tuition to copy nature; and I am still at it. That is why an artist has no reason to balk at copying a simple, though very beautiful, insect. I asked six amateur fly-dressers to make artificials from my colored drawings of the natural insects (which I had taken the greatest pains to depict in the attitude they most generally assume). All six declined to do it. They said, "We must have an artificial fly for a pattern; not a picture."

So to learn the art of fly-dressing, I suggest the amateur to do as I did: find out by practical experience the most effective way, the quickest and easiest way. Most of the amateurs claim their method the only right one. Yet some tie on wings first and body afterward. I found it much more simple and convenient to make the body first, then tie on the wings; and finish up by putting on the hackle. I have previously stated that there are no rockbound rules; each one works out, or blunders out, his own method.

Roughly speaking, there are three kinds of bodies: a fat body that needs a foundation to thicken it; a thin body wherein the material is wound simply around the hook; and a detached body, when the tail is made separate from, or raised above, the body yet is part of it.

Before giving a description of the making of bodies, which will be understood better from the diagram sketches, I deem it wise to mention a few important features concerning the method employed.

It is most important that the tying silk be well waxed, so that it will not slip, but will hold fast and solid when you tie up the tinsel, raffia or any other material at the head or tail.

If the body is light in color, it requires a light, delicate yellow or pale blue tying silk; and it should be waxed with transparent wax that will not discolor the silk. If the body is to be dark, and dark silk is used for tying, the black shoemaker's wax is the best.

If you get used to working without a vise the thumb and first finger of the left hand must be trained to grasp, firm and tight, the hook, the body, and the wings. This would seem a trivial matter; but at first you will constantly be dropping and slipping the fly unless this firm grasp is attained. The thumb and forefinger nails of both hands should be long enough to pick up readily small single hooks, then bristles, and delicate wisks used for the tails. I have said this before; but the matter is so

THE MAKING OF AN ARTIFICIAL FLY

important that a repetition is not wasted. You will save a good deal of trouble by discarding the use of tweezers every time you pick up objects.

The tying silk must always be wound and pulled as tight as you dare without having a break. Especially does the end tie need to be pulled perfectly tight, that the fly may not come apart or the wings come off.

For the fastening-off knot I use the two turns (perhaps three) and half hitch. I know it is the old-fashioned way, but I am used to it, and it is so simple that I see no valid reason to tie otherwise. Mr. Halford severely condemns it. He favors the "whip" finish, which is more trouble and gives a lot of bother, especially in tying on the hackle. If the half hitch is drawn good and tight, and the finish is waxed and varnished, the knot stays secure. However, I give here the whip finish as described by Mr. Halford, so that those who desire to do so may use it. The whip finish for tying off is made thus:

"Lay the end of the tying silk back towards the tail to form an open loop, and work one turn of the loop round the neck of the eye. Similarly, work three more turns of the loop, passing it at each turn over the eye. Holding the hook and turns of silk firmly between the left thumb and forefinger, draw the end of the tying silk down with the right hand until the knot is quite tight. It is essential in this operation to proceed slowly, so as to allow the warmth of the finger and thumb to soften the wax, and allow the silk to draw freely. Cut off the remnant of the silk, varnish the knot thoroughly, and if in this operation the eye is filled with varnish, do not neglect to clear it."

It will be advisable to state here, for the benefit of those not used to eyed hooks, that these directions apply exclusively to eyed hooks, and not to plain hooks lashed or tied on gut. Artificials tied on eyed hook are in every way superior. They can be tied on the gut and taken off for a change of fly at any time desired. If properly tied they are sure to wear better and give more satisfaction in every way.

THE WAY TO TIE A FLY

You can much more readily grasp the method of making a fly from diagram sketches than from a long, detailed description, which would only prove wearisome and confusing.

The amateur must be content to master first the simple, easy flies with upright wings and flat bodies, like the commercial flies. Those bodies of different thicknesses—that is, fat in the middle or tapering down thin—are best left till the plain, even body is understood.

After you are past the rudimentary stage you can then make efforts to copy the natural insects, no two bodies of which are alike in thickness. Then, afterward, if you wish to copy beetles, spiders, palmers, or caterpillars, with legs all along the body, try your hand at it. And if you want to make a split-wing fly, now widely used and by English experts considered best for floating flies, all you have to do is to wind the silk between the wings instead of around them. I do not think these split-wing flies are good, because the natural up-wing drakes (comprising over forty species) all float on the water with wings tightly closed and not flat. Another thing, with the aid of oil, the tight-wing fly will float upright just as well as the spread-wing.

Halford's book is nearly twenty years old. Leonard West is more true to nature; while the drawings are amateurish, they do make a beginning to lead the way to some progress in the art of fly-making. Therefore it remains evident that three ways of tying wings should be understood: the upright-wing flies to represent drakes; the down- or side-wing flies for the duns; and spread-wings for spinners. Spiders, beetles, and other insects less in use are but a variation; and, after all, I doubt if they are worth the trouble of making when drakes and duns are on the wing.

To make a fat body, you first wind a foundation. For this I use mercerized cotton, which is soft and silky, yet not bulky, and is much cheaper than silk. Remember, all winding must be carefully done, smooth and well laid, not lumpy or irregular.

With the winding silk you tie fast the ribbing or tinsel and body material at the bend of the hook, or tail end, and the thorax or shoulder material you tie at the eye end. The thorax should never be too thick, because the tying on of wings and afterward the hackle makes the thorax thicker still.

A thin body requires the material tied round the hook itself, whether it be of silk, raffia, horsehair, quill, or tinsel, or a combination of any of these.

To wing the fly, after the body is completed, and all ends neatly cut away, you must first prepare your wings. Cut out the tips from feathers of small birds' wings, or neatly cut from the sides of large wing feathers; then trim these to the right size, having them evenly matched, with a corresponding feather from each wing of the same bird cut down to the appropriate size. When this is done, with the left finger and thumb hold the wings tightly in place and tie fast. If the wings get out of place, too far forward or backward, force them exactly right before tying off. After it is tied secure, cut off the stumps as neatly as possible and the fly is then ready for the hackle or legs.

Before you tie on the hackle it must, like the wings, be prepared. First, strip off the downy plumes at the base of the quill. Then, taking the extreme point between the thumb and forefinger of the left hand, with the right thumb and forefinger slightly moistened, stroke back the whole of the

THE MAKING OF AN ARTIFICIAL FLY

plume, except the small tip portion, which should be held tight to the body of the fly. Then wind the tip portion in position with winding silk, taking two turns and a half hitch. For extra safety, tighten the tip portion with the left thumb. Then grasp the quill end with the right thumb and finger (or with the tweezers, if you use it) and wind the plume round in front of the wings, then back of the wings, and again as many times as necessary.

Some hackles are required to be scant; others are bushy, long and full. In winding the hackle feather round, take care to have the glossy side in front to face the eye of the hook.

There are times when two small hackles are better than a large one, if the legs are required to be quite short. Long fibered hackles cut short are too stumpy and blunt.

During the winding process the hackle plume must never leave the fingers or pliers; for if left to itself it will unwind like a spring. When the last turn is made, force the quill end of the plume by the little finger under the body toward the bend of the hook; then secure tightly with winding silk by means of two or more turns and half hitch.

If the tie-off seems to be not quite secure, and a stiff plume quill is more needful for a stout tie-off, it is wise to put on a piece of wax the size of a pinhead over the tie. Then varnish, using a fine-pointed sable-brush, as used by artists.

To tie on wings for a dun or flat-winged fly, the pair of wings, after being made the right shape and size, should be evenly placed on top of thorax or shoulders, with stumps over the eyes. If the wings are wanted to lie at the sides, arrange them so that they lie farther down the side of the body; then hold them firm and flat while you fasten them secure with tying silk, as before described.

To make a detached body: I make this separate from the hook, tying the material (be it quill, horsehair, or silk) round a piece of boar's bristle, double or single, according to thickness of body required. I sometimes use up quills of small-sized feathers stripped clean. These small quills are not quite so firm or solid, but they suffice. After the body is wound and tied it is quite easy to fasten it to the hook at any angle desired. Sometimes I wish to make the tail erect from the hook a short distance: I then work in a bristle by overwinding from the thorax.

Though I have left the wisks or tail fibers to be mentioned last, they should always be fastened on first, under all other materials, whether they are to be at the head, for horns, or at the tail. It is very necessary that wisks be correct as to number, size, and color. The right wisks, I am sure, make a great difference in your fly as to deceiving a trout.

I am a strong advocate of the use of tinsel. It is used in nearly all my new patterns, because the un-

THE MAKING OF AN ARTIFICIAL FLY

der bodies of all flies are very light, either silvery or golden in color. The glint of bright tinsel must attract attention.

What I desire most is to impress upon anglers that to tie their own flies enhances tenfold the delight in their craft. I don't think money is saved; it is far cheaper to buy the commercial flies offered by the tackle men than to provide the materials, including the hooks, to say nothing of the valuable time spent in making the flies. But there is another side to the question: the fun of killing a trout with a fly tied by yourself from the natural insect is an achievement far more satisfactory than catching a fish with an imported or domestic fly made by other hands. The ardent fisherman, wise and expert, knows by the time and season just what flies are hatching out and on the wing; if he makes his own flies, he provides accordingly.

XIII

A TEST OF THE NEW FLIES

If the angler will take pains to notice he will perceive that I have used great care to depict each of the varied insects in its most characteristic attitude when it alights on the water's surface or on some other object, so that he may more readily identify each specimen and thus become more familiar with trout insects. Had I shown them in spread-wing fashion, like that familiar in scientific books on entomology, the different species of each class would be so much alike as to be impossible of identification.

If we carefully observe some of the drakes as they lie at rest underneath a smooth rock at the water's edge—always at the side on which the sun is shining and opposite to where the cold wind blows—it will be seen that the wings, feet, and cocked tail never

A TEST OF THE NEW FLIES

vary or change, as do those of land insects. The wings never separate, or lie apart; the tail never loses its jaunty, upright appearance; and the feet assume the same position while resting. This is true also of browns and duns: they never lift the wings up, nor spread them out, but always their wings lie flat on the body when they are not in flight.

The test of my new patterns that have been made are most gratifying to me personally; and I sincerely trust that anglers in various localities will select just a few of the copies of the natural insects and try them; I care not upon what conditions, though the result will be much more pleasing to all concerned if the flies are used at the specified time given in the plate charts. Even if tried otherwise —that is, any fly cast at any time—trout will, I think, be induced to rise sooner than with the best fancy fly.

It would be a very different thing were I endeavoring to urge upon my brother anglers any new fancy flies, new inventions, unusual or freak flies. I see this being done every season, much to their disadvantage and my own. But the wise angler, I hope, will see the object of this book: to make anglers for trout follow a scientific method of the highest importance whereby we attain the best there is in angling.

When this work was begun several years ago, entirely for my own private benefit, I saw in it great

AMERICAN TROUT-STREAM INSECTS

possibilities for good if the method—going back to nature—could be adopted universally. If the method does not become universally popular, I have lost nothing; I shall be the sole, yes, solitary, gainer.

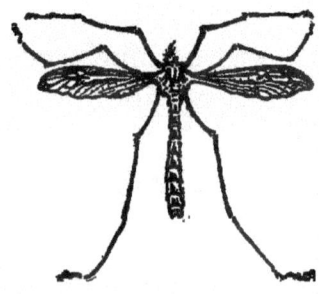

PART II
NEW ARTIFICIAL NATURE LURES

INTRODUCTORY NOTE

CONCERNING THE RAPIDLY DIMINISHING NATURAL FOOD OF FRESH-WATER GAME FISHES

It is interesting to note that the anglers' so-called game fishes are all edible, and the gamiest of all—trout and salmon—are the most highly prized as food for man. For that reason alone, they should be encouraged to a larger growth and greater abundance.

In Dame Nature's bountiful supply of food for man, it is fish, above all others, that live to eat. From babyhood to adult stage their only object in life is eating. Other creatures—animals, birds, and insects—devote part of their time to trimming fur and feathers, building abodes, and maternity, with many other duties necessary to their well-being. It is not so with fishes; their entire time is devoted to the sole object of getting food—with one or two exceptions. So it is undeniable that all fish are wont to haunt places where food can be ob-

NEW ARTIFICIAL NATURE LURES

tained in the greatest abundance, and that they grow more quickly, attaining greater size, in such localities.

If you take two six-inch trout in the spring, place one in a mountain brook where food is always limited, and the other in a large river where various kinds of food are usually plentiful, by fall the river trout will have gained half a pound, while the trout in the brook gains but an ounce. It is precisely the same as fattening hogs and feeding up chickens for market.

If you feed a trout upon artificial food—liver, chopped meats, etc.—the effect is apparent in its lack of gamy resistance during capture on fly or bait, and in the taste of its flesh when consumed as food for man. It is as "water unto wine," compared with the wild trout fed by nature's bounty.

The natural food of game fishes is quite varied, and cannot be propagated by artificial means. If the supply of food be much curtailed, game fish move—if they can—to new pastures; or, when confined to a given space, they will eat each other. They are very acute in their search and capture of food, and never seem to be satiated or satisfied. Both trout and bluefish, gorged nearly to suffocation, readily respond to a lure.

The natural food of fresh-water game fishes, valuable according to the order named, are: insects that fly above and on the water; their larvæ, that live and

INTRODUCTORY NOTE

creep on the bottom, to rise through the water to the surface to attain maturity; a variety of minnows and young fish food; crawfish, helgramites, lamper eels, frogs, grasshoppers and caterpillars.

It will be seen that if man permits nature to work its own way, these creatures, by feeding on each other (though in some instances multiplying in vast numbers), are kept down to a reasonable extent, and the balance of living things is about evened up. It is rare indeed in our day to find a glut, or an overproduction of any one species of fish, or fish food, like we used to observe in days gone by. Numerous reasons may be cited to explain this, but the principal reason is a decided shrinkage of fish food, these live baits being captured in great quantities for the ever-increasing army of anglers. Ten per cent., possibly, are consumed by game fish; the rest destroyed. Similar conditions prevail with sea food used in marine fishing.

Fresh-water fish food can now be procured only in limited quantities. Minnows and frogs are protected—as they should be—by stringent laws.

In times past live minnows were ten cents a score; now they are fifty cents, and not always to be got; indeed, they are in places difficult to procure at any price. Thirty crawfish are a fair day's bait for bass fishing; after a few days' hunt, the brook from which you get them becomes wofully scarce if other anglers seek supplies from the same place. In

NEW ARTIFICIAL NATURE LURES

olden times we could soon dig a day's supply of lampher eels from the sand-bars at the bend of the river where they abide; but the constant digging for them by numerous anglers keeps the supply at a very low ebb, and it is now useless to spend the time.

These conditions will not improve in the future; they will rapidly grow worse. The time is not far off when the securing of live bait for game fishes will be an utter impossibility; when fishes will be stunted in growth, their supply curtailed, in some cases absolutely cut off. Of course, this is a big country—changes are slow; but they are sure.

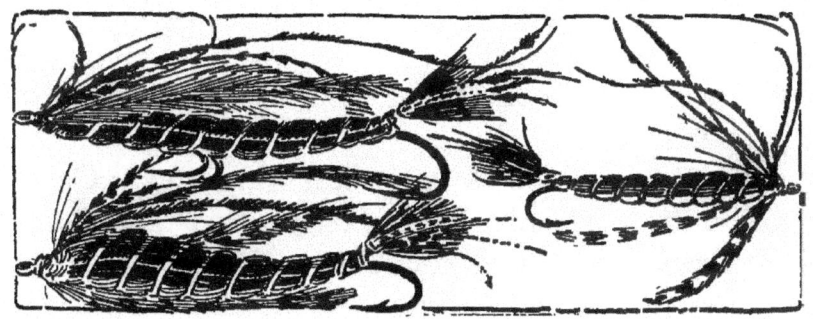

XIV

NEW LURES THAT ARE TRUE TO LIFE

It was originally intended to issue trout-stream insects in book form a year ago, but the articles contributed to *Forest and Stream* on nature lures received such a wide and cordial approval from a large number of anglers all over the United States and Canada that it was thought wise to incorporate the two subjects in one volume, especially as many anglers practise both lure and fly fishing. Indeed, the writer's desire is to induce anglers to fish with a sane lure as they would a fly, in order to raise that method of fishing to a higher plane.

To that end, a systematic study was begun of what game fish consume as food, and careful pictures were made in colors of the living creatures. The next step was to invent and construct the best possible imitations of them with the limited materials at command, so that they might supersede all others heretofore done. The following list is

NEW ARTIFICIAL NATURE LURES

the result (each one to be described in detail in its proper place):

Floating minnows with quivering fins. Crawfish with a pliable tail. Helgramite that floats, with movable legs. Frogs that both float and swim. Grasshoppers that can be made to skip on the water. Caterpillars that float. Dragon-flies used as a dry fly.

FEATHER MINNOWS FOR BASS, PIKE AND TROUT

My first efforts after the completion of the work, "Natural Trout Insects and Their Artificial Imitations," was to make a new minnow, to be cast like a fly and reeled in like a minnow. I then determined, for my own satisfaction, to continue in the effort to get something new, and true to nature, in the shape of a feather minnow for bass—one weighted for bottom fishing, and one light for surface fishing. In recent years it has been my impression that minnows (certainly the best all-round natural bait for game fishes) have not yet been perfectly imitated so as to be a universal success for either bass or trout.

A hard substance, like wood, metal, or quill, surrounded by a fierce array of treble hooks, would not allure me, were I a bass or a trout, in place of a soft, fleshy young fish. Before describing the feather lure, I must explain that in the trials I have made of the artificial minnows now on the market they

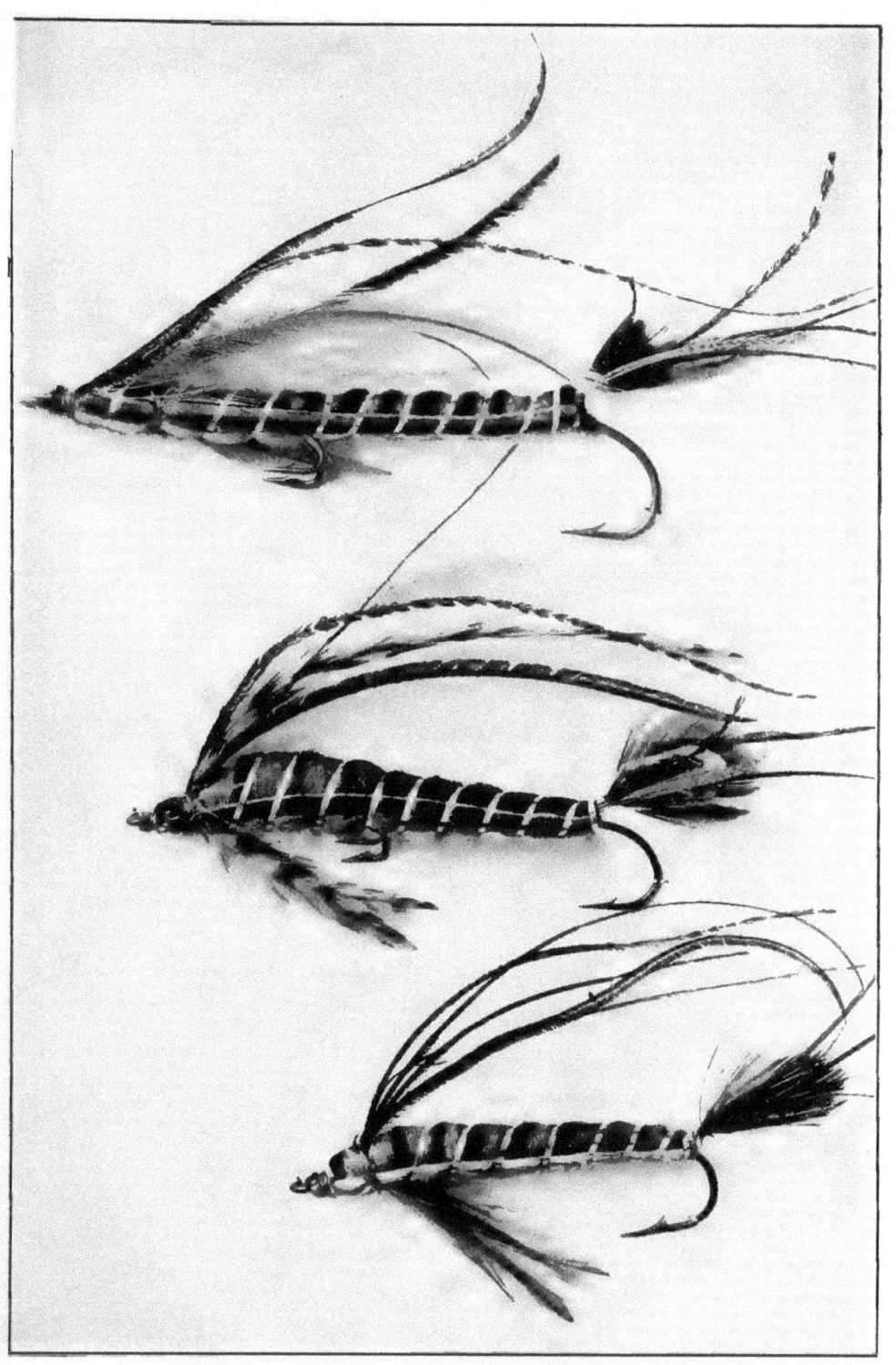

FEATHER MINNOWS FOR BASS, PIKE AND TROUT

are not successful in the waters I favor—that is, a deep sluggish river with rocky bottom, like the lower Beaverkill and the upper Delaware round about East Branch.

I have repeatedly seen bass follow up various artificial lures for some distance without making a strike. The same thing happens, at times, with the painted minnows with metal fins to make them spin. On the other hand, with a supply of live minnows, crawfish, or lampers, one can go to a pool and often capture a dozen bass that will average over two pounds' weight, almost any time of day in late July, August and September. But live bait in sufficient quantities is most difficult to procure.

This non-success of hooking a fish is not due to my inability in casting a lure. I can cast a lure, as I do a fly, dry or wet, quite delicately and far enough to catch bass or trout. Others have had the same experience. I met a man wading the stream, casting a wooden painted lure of well-known make, and I could see by the way he cast that he was expert at the game. In answer to my question, "What luck?" he said that he had not got a single strike. I took him to a sand-bar nearby, and a conveniently hidden spade, and together we dug a few lamper eels. He immediately caught a three-pound bass. "Now," said he; "I know bass are here."

I have no desire to try to persuade thousands of

NEW ARTIFICIAL NATURE LURES

bass fishermen from the use of wooden or metal lures. The new ideas I propose are primarily for my own use and pleasure; but I offer them, after proving them good, to my brother anglers everywhere, to accept or decline as they choose.

Beyond question there are a host of anglers, like myself, sick and tired of testing new devices and ideas that lately flood the market claiming under various pretenses to be perfect killers. I do not deny their value or usefulness under certain conditions and methods of fishing. The aim and object of my work is to imitate both flies and lures as nearly as possible like the natural food game fishes consume. This principle, everybody must agree, is a good and right one.

In describing this feather minnow, the larger of the two is selected from a great many different kinds I have made. The body is soft, though solid, and wound over a long, powerful single hook, which curves under the tail, making the hook partly invisible from below. A double hook, smaller in size, is placed under the fore part of the belly, hidden by two pectoral fins of speckled cock's hackle. In my first efforts I placed the double hook on top, near the tail, hidden by a stiff dorsal fin of turkey-wing feather, cut to shape. But this, I found, would interfere with the minnow's floating belly down—a position absolutely necessary in the water. Although this would be a more effective hold when the

NEW LURES THAT ARE TRUE TO LIFE

fish grabbed it, and looked more natural as a minnow, I was forced to place the double hook underneath to get the upright attitude while moving in the water.

The upper back and head are colored in alternate stripes of black, green, and brown. The tail, made of the feather tip of turkey's tail, is of bright, metallic luster, edged in white. Graduating from the back along the sides are two stripes of green and blue which fade into a perfectly white belly. The whole is wound in silver to imitate the shiny scales. This larger size minnow is weighted to fish sunk in deep water, though not too heavy to be cast with a fly-rod.

The smaller size minnow, No. 2, is like No. 1 in color; but the body, being made of cork, is somewhat wide at the shoulders. The bodies, well padded to stand being chewed by the fish, are made of colored wool and mohair. A feature of great value in this minnow is that, while solid enough, it is sufficiently soft for the teeth of the bass to prevent his being scared as he would be on such a hard substance as wood or metal, which is the chief objection to be found in those lures.

No. 3 feather minnow is smaller and is light enough to be cast like a fly and reeled in like a minnow. You can swim it near the surface in shallow water. You can cast it down a swift runway and swim it back to any depth, according to the speed you reel. You can weight it to cast out to the bottom of deep pools and reel slowly in; or you can use it as it is at the surface of a pool. No matter if the game be bass or trout, pickerel or pike, wall-eye or perch, big dace or chub, it is equally serviceable for all.

A few of the good points I claim for these minnows are:

(1) Fidelity in size, form, and color to a real, live minnow.

(2) Hooks so placed as to hold fast, without scaring away the fish.

(3) Fish find a similarity in touch to the flesh of a minnow.

(4) Light enough to be cast like a fly with trout-rod and tackle.

(5) Most desirable of all, fish will strike it in plain view of the angler, as they do at a fly.

XV

SHINY DEVILS

GOLD- AND SILVER-BODIED FANCY MINNOWS FOR SALMON, BIG TROUT, AND BASS

IN studying these minnows from every standpoint and developing them so as to make, as perfectly as can be, practical, all-round sensible lures for our native game fishes, and following my efforts to produce an exact copy of a natural minnow in a soft yet solid material, I conceived the idea of making some "fancy" minnows—shiny devils, by name—where no effort is made to imitate the minnow except in the under body or belly. This is contrary to my often expressed, determined belief that it is best in every instance to copy nature.

I do it for this reason: After testing them side by side with imitation minnows (both floating and sunken) I shall have tangible proof which of them game fish most prefer; though I imagine that, while

NEW ARTIFICIAL NATURE LURES

these fancy minnows look more attractive and beautiful to our eyes, the natural imitation will win out, just as nature flies have done and will do over fancy flies.

Seven years ago I wrote an illustrated magazine article with the title, "Try Bass and Trout Flies with Metal Bodies." It described how I invented four flies, two with gold and two with silver bodies, having black, brown, gray, and white wings. Recently there appeared a magazine article on dry flies, in which the writer stated, "These metal-body flies are still in use and are far superior to old favorites." Personally, I have discarded them for my more recent "nature flies," which I consider are as far as can be got in the right direction.

When I showed the feather minnow to a member of a well-known tackle firm he said, "There is nothing new under the sun," and placed before me a bass minnow (tarnished with age) made and used forty years ago; thus proving my often-repeated statement that we have gone backward in the making of lures, so far as nature is concerned. This old minnow has a tail of peacock's harl, with stripes along the back of red, black, and green wool. The belly is common tinfoil, the whole wound over with silver tinsel. I asked permission to copy it, and the result is seen in No. 1 of the six shown on the page of illustrations for this chapter. The only alteration I made was the additional plume, giving a more

GOLD AND SILVER-BODIED FANCY MINNOWS FOR SALMON, BIG TROUT AND BASS

shapely body. With the exception of a solid silver body, this shiny devil is the same, or nearly so, as the smaller sized feather minnow with white under body. This shiny devil has proved exceptionally good, better even than the feather minnow because of the brilliant solid silver body.

In No. 2 the entire head and body is wound with silver twist and tinsel, the tail is a speckled guinea fowl's feather, and the plume is of green harl and wood duck tips.

No. 3 has a solid gold upper and under body, the tail is a deep orange breast feather of ruffed grouse tipped in white.

No. 4 has a solid silver body, with a tail of white downy feather of the loon, and a very bushy plume of mixed feathers.

No. 5 is a smaller devil, with solid silver body and green head. The plume is made of the wing feather tip of the quail, and the tail is made of tail feathers of ruffed grouse.

No. 6 is the largest—a shining mass of gold from head to tail, except the eye, which is green. The plume is the beautiful chocolate and orange feather of Egyptian quail, with a red cock's hackle for tail.

Even a detailed description fails to give one-half the beauty of color in these minnows.

In testing some of these unusual and original lures I have experienced some rare good sport with three species of game fish, viz.: trout, bass, and wall-

NEW ARTIFICIAL NATURE LURES

eye. In a later chapter I shall describe the successful tests—not so frequent nor so thorough as I hope them to be another season, nevertheless sufficient to convince anglers that these minnows are worthy to rank as lasting lures.

When dry, the shiny devils appear somewhat similar to the regulation salmon flies; but it must be remembered that these plumes, when wet, cling close to the back of the lure, thus forming a dark varicolored upper body that resembles a minnow far better wet than dry.

It is well known that two of the most deadly materials in use by fly-makers are shining metal and peacock's harl; in fact, upon some English rivers, the "Alexandra fly" and similar lures made from these materials are not allowed in fishing, because of the supposed "unfair" dead sure killing.

In the manufacture of flies, minnows, and other lures, many changes have been made during the last forty years; but, to my mind, they have not been improvements upon the old and better lures. I believe the tendency of modern lures is in the wrong direction. Highly polished, vividly colored creatures, of a hard, machine-made material, describing all kinds of acrobatic water stunts, are directly opposite to and lack the artistic handiwork of those beautiful lures which in bygone days gave better results, I am told by well-known expert bass fishermen of fifty years' experience.

SHINY DEVILS

Goethe says, "Encourage the beautiful; the useful will take care of itself."

My theory is that the useful lure is the one which gets the fish to imagine it is taking its natural food and grasps hold when the fish strikes. In other words, a useful lure, to be a good one, must be made upon the basic principle of an imitation, in some way or other, of the natural food of the fish, gamy or otherwise.

Of course, if these soft metal-body devils are found by other anglers to be more effective for better sport and larger fish than is possible with the prevailing plug lures arranged in many barbed hooks, it will be a great gain in many ways.

To be candid, there is one serious drawback to these shining lures, viz.: a tendency of the metal to tarnish and lose its brightness after being wet. For this I can find no remedy except to dry them carefully after being used, and to keep them wrapped up from light and air. I have tried the best imported metal twist and tinsel; and it is no advantage to use pure silver, as it is the most quickly tarnished. The tinsel is made much brighter, when tarnished, by the careful application of a little metal polish. It is quite different with the feathers; they are all natural and undyed, and both their form and color are retained after very hard, rough usage.

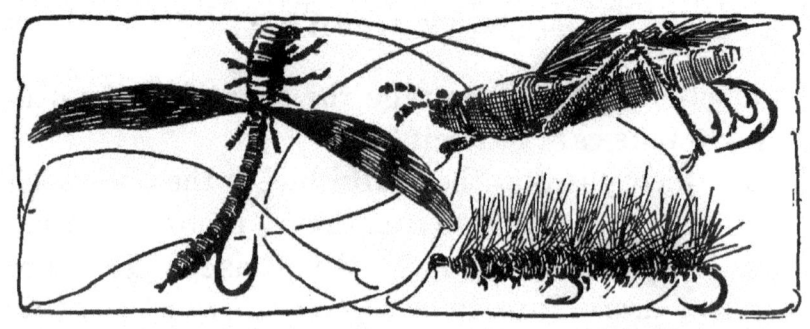

XVI

NATURE LURES FOR SUMMER FISHING

IMITATIONS OF MINNOWS, GRASSHOPPERS, DRAGON-FLIES AND CATERPILLARS

In the course of my study and development of minnows, it is natural that a seeker after truth and perfection should make some advances. But in time the improvements get to a stage whereby accurate imitation, as to both form, and action in the water, is attained. This I believe will be found in my quivering-fin minnows shown in the accompanying photograph.

It would be madness for even the very foolish to deny that game fish seek localities where food is most abundant; also, to deny that fish will and do prefer lures and baits like the regular food they are wont to eat. My theory is, give a fish the food it likes; if that is not possible, then offer an imitation of that food, as nearly as it can be made with the materials at command.

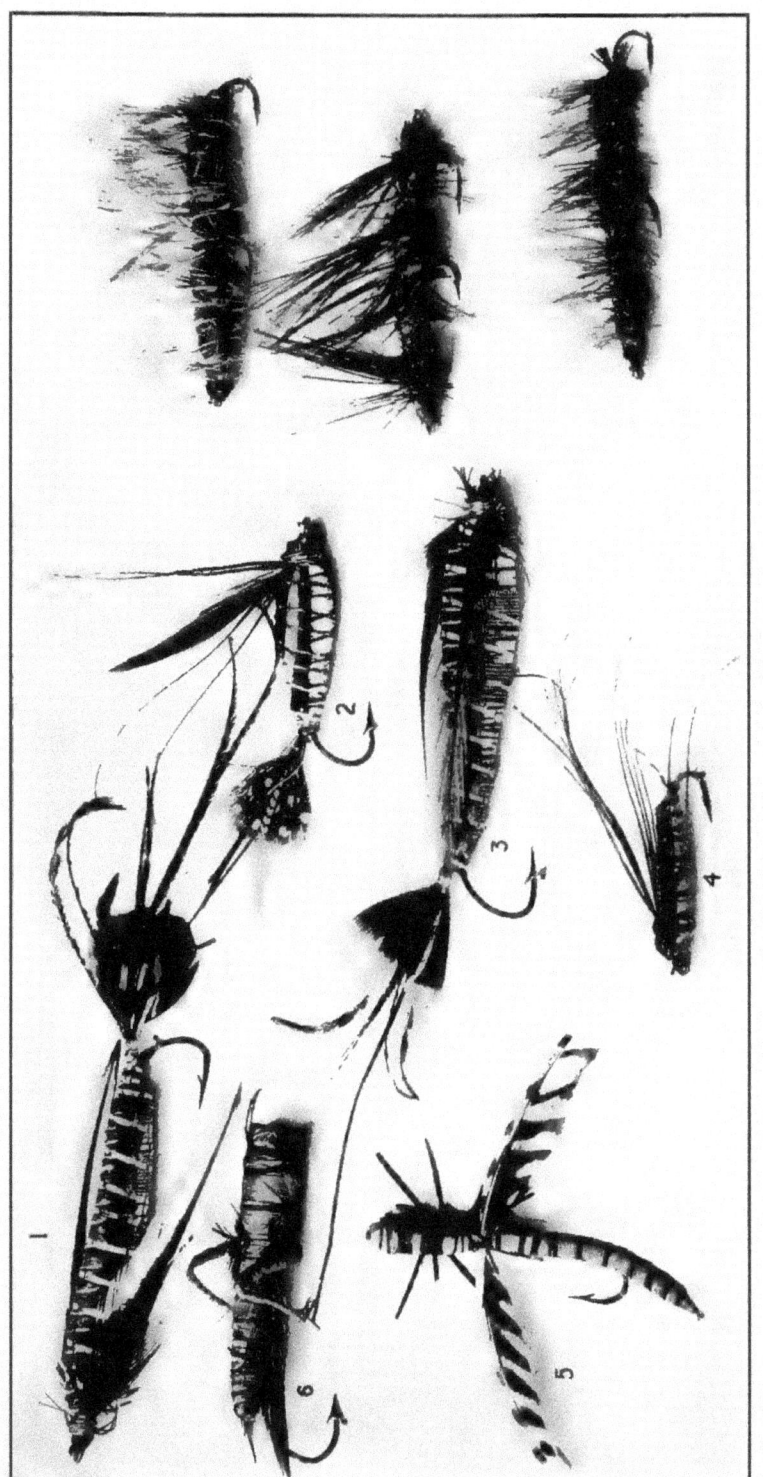

NATURE LURES FOR SUMMER FISHING
Imitations of Minnows, Grasshoppers, Dragon Fly and Caterpillars

NATURE LURES FOR SUMMER FISHING

This I am endeavoring to do to the best of my ability. An artist is particularly favored in that he can (with more or less success) copy nature, and after patient effort produce what less-favored persons would scarce attempt.

All rivers or bodies of water contain certain kinds of fish food that preponderate over others. Lakes of deep water and few weeds have an abundance of minnows and few frogs. Shallow lakes, with plenty of weeds, grass, lilies, and scum, breed frogs, dragon-flies, grasshoppers, and much other surface food. Rivers that have muddy and sandy bottoms breed entirely different insects from those rivers with rocky or pebbly bottoms. And, whatever the water conditions are, the food varies considerably according to season. Just as the summer's heat gets powerful, aquatic insects become smaller and scarcer. Then nature supplies certain fish food of a larger kind. After June the water is thick with the larvæ of various big stone-flies and dragon-flies; the land in close proximity to the water is fairly alive with grasshoppers (winged and wingless), and the river and lakeside trees are loaded with different species of caterpillars. It is quite true that from early spring to late fall, minnows certainly do form the major part of fish diet. But minnows are sharp enough to haunt the shallows where big fish fear to go. It is only during the night time in summer that large fish prowl around the sides to get them.

NEW ARTIFICIAL NATURE LURES

The minnows here shown are in some respects an advance on the feather minnows previously described. They are much harder to make, because the two parts are separately constructed, the back of cork, and the belly of wood. They are so made that they float upright and require less movement in playing; in fact, fish have taken them while perfectly still in repose. I discarded the plume, which forms a prominent feature of the feather minnow, and used a pair of quivering feathers placed at the forward sides of the body as pectoral fins.

This minnow will be found most effective when cast downstream, then gradually reeled back from side to side of the river, covering as much water as possible.

I am confident no minnow heretofore invented is so deadly as anglers will find this lure to be if fished in the method I describe. My sole object has been to study and investigate how to get the best sport, the largest fish—which are always hardest to capture—in the highest and most skilful manner, and to supply the means. In order to accomplish this I have during the last three years made careful color pictures of every kind of food that game fish eat, and I now have what I hope will be considered by anglers a complete line of nature lures, as perfect in appearance and action as it is possible to make by hand.

A live grasshopper that floats downstream in the

NATURE LURES FOR SUMMER FISHING

daytime has little chance to go far before it is seized by the first large fish in sight. If you hook one and float it alive—nay, if you do so half a dozen times—you will see how cleverly the fish nip the grasshopper off without touching the hook. This smart trick is not so easy with my nature lure grasshopper, which is made to float standing upright on the surface. If you cast it out where trout lie, either on a clear, glassy surface, or down a rippling runway, you should use a fine six-foot gut leader strong enough to hold a bass or trout. Attach the leader neatly to the bait without any additional feature—spoons, sinkers or other attractions—playing the rod-tip so that the bait skips along in short jumps, to imitate the natural insect when by accident it falls on the water. Strike instantly the bait is taken; for the fish can immediately tell the difference between artificial and live bait.

The advantage of grasshopper fishing is that none but large fish will go for it. Also, you may fish with fairly good success on hot, sultry days, when flies are not so effective; though evening fishing is, as always, the best, because both trout and bass are at that time more active in their search for food.

The dragon-fly, of which there are two sizes made, will be found best to use as an ordinary fly. Cast it out on the surface to float a while, whip it off again without being wetted, and keep repeating the cast to different places. This fly is so light that it

NEW ARTIFICIAL NATURE LURES

can easily be cast among the weeds and lily pads, where it will be taken greedily by large perch, pickerel, bass, and pike. For river fishing, almost any plan will induce a rise where fish happen to be.

Modern anglers do not, I believe, take a fancy to caterpillar fishing, probably because there are no good imitations. They do use the brown and black hackle fly of various sizes, these being supposed to imitate the brown and black hairy caterpillars so abundant in early spring and summer. Seventy years ago, in the time of Ronalds, excellent caterpillar imitations were made and used; not in the shape of a fly, but copied with fidelity and truth from the natural insect. The samples here shown are brown, black, and gray, the two former being most plentiful in the spring and the latter in late summer and fall. They should, like the grasshopper, be attached to a long fine leader, and floated leisurely along the surface under which trout may lie.

A short description, including the colors, of the varied lures here illustrated will give the angler a better idea of their value.

No. 1 is a copy of the silver shiner or dace. The belly is of solid silver, being wound alternately with silver twist and tinsel. Along the sides is a strip of vivid green and blue, which tones down after being wet. The back is olive green with black stripes. The quivering side pectoral fins are black and deep

NATURE LURES FOR SUMMER FISHING

yellow. This feature forms an entirely new idea in artificial lures. However tranquil the water may be, these fins have a gentle waving motion that is very lifelike and attractive. This is a large minnow, and, as with No. 3, it will be found equally killing for big brown and other trout, bass, pike, and salmon.

No. 2 imitates a small red-sided minnow familiar to fishermen as the red stickleback. It has a white belly, red sides, and green back. It has killed, and is highly suitable for, good-sized brook trout and rainbows.

No. 3 is similar in size and shape to No. 1, and is a copy of the redfin minnow. The belly is shining gold tinsel, the quivering fins are red and black, and the back is of the same color, with a stripe of pale orange and green running along the sides.

No. 4 is a very small minnow with white belly and olive back, and is intended to represent the young of various species of larger fish.

No. 5 is a dragon-fly, with a solid body of cork re-enforced with strong silver wire and wound tightly with green raffia. The wings are feathers of snipe. The exceeding lightness in weight makes it possible to cast like the usual large-sized fly.

No. 6 is the June green grasshopper, made of solid cork wound in vivid green raffia. It floats upright, and the quivering back wings of red cock's hackle make it a choice irresistible lure.

NEW ARTIFICIAL NATURE LURES

From the foregoing chapters it will be seen that I have made a large variety of minnow baits—as to size, shape and color. Not more, I believe, than is warranted by so important a part as the minnow takes in fish diet; for every angler will agree that the minnow is undoubtedly the most popular lure for all game fish, under any condition, or in any location or season. As the seasons come and go, we find live minnows scarcer, more difficult to procure, and the time is near—in fact, it is here now—when a good substitute is of the greatest value; for game fish need all the food that nature now provides, to attain large growth and still be abundant.

XVII

ARTIFICIAL FROGS THAT WIGGLE THEIR LEGS AND FLOAT

For years I have vainly tried to get a fish strike on the various imitation lures, most of them made of rubber. They are not only miserably poor copies of nature, but, from their weight and clumsiness, they act in the water as dead, inanimate things. No matter how skilfully they be played, trout and bass take not the slightest notice of them. I would as soon fish with a "tooth brush" at the end of my leader; there would be more chances of a strike. Most expert anglers will surely agree with me in this after one trial of them. Particularly so of the painted rubber frogs, grasshoppers, worms, and other imitations intended to replace live bait.

This does not refer to "plug lures," which are not, I believe, intended to imitate any living thing.

To the end that something may be available for anglers without their having to use live bait (so hard to get, so hard to keep fit for use) I have spent

NEW ARTIFICIAL NATURE LURES

considerable time in study and experiments to get at *just the perfect* nature lure that will appear and act as enticing as the natural food does to bass, trout, and other game fishes.

A thoroughly good and useful article is not usually dreamed of over night, and completed the following day. I made twenty-three different models of this jumping, floating frog, before I reached the desired result. Many hoped-for improvements were discarded because of some undesirable feature. These frogs and other lures are the result of continuous effort in practical trials and experiments in order to gain three important points heretofore not accomplished and demonstrate their superior value as lures.

First: They must be light enough in weight to cast with a fly-rod, and to float upright and naturally in quiet or turbulent water, yet strong enough to be chewed and yet not destroyed.

Second: They must be soft to the touch, without scaring the fish.

Third: They must have perfection in form and color, combined with a natural action similar to live bait when floating in the water.

For these and other reasons, I determined to get for my own use and satisfaction, and to give the true angling sportsman, a lure *to lure* and not to scare. Of this I am convinced: if the present style of bass lures continues to develop, in a few years'

ARTIFICIAL FROGS

LAMPER EEL AND HELGRAMITE

ARTIFICIAL FROGS

time, Mr. Bass, likewise Miss Trout (both wise, alert, and discerning fish), will absolutely refuse chunks of wood, rubber, or metal, of whatever shape or however finely finished. They know as well as we know, that such lures, tearing through the water by them, are not food. Their action in taking them is merely antagonistic caprice; the lures in time will cease to annoy and therefore utterly fail.

This reasoning is sound; for when we miss a strike, I know that we never get the same bass to go for the lure a second time. When we cast again, if taken, it is sure to be another bass. It is altogether different with live bait. When a real minnow, crab, or frog is gorged, the fish is ready and willing for more; the effect in its stomach is obvious and most pleasing. It is very natural to suppose that Mr. Bass is quite satisfied with but one trial at a piece of wood. The dullest, most stupid animal in creation would undoubtedly remember such a base fraud. Bass are neither dull nor stupid, but, as Dr. Henshall rightly says, "the gamest fish that swims."

Furthermore, to each of these pieces of wood is attached from three to five treble hooks. It seems impossible to imagine a true fisherman could be found to face his gamy antagonist without a blush of shame while using fifteen barbed hooks on a single lure. The success of these lures must be very doubtful from the fact that new and different pat-

NEW ARTIFICIAL NATURE LURES

terns succeed each other every season. If a lure is good one season, it should be so for all time.

The majority of black bass invariably prefer to abide near the bottom, in water from four to twenty feet deep. It is round the shallow edges of rocks, sand-bars, and edges of lakes where they congregate. They lie still most of the time, like other game fishes, to pounce periodically upon passing prey during the daytime; then at night they swim about the shallows foraging for food. They will follow a lure some distance before they grab it; in fact, they often follow a lure within two feet of the boat, making a grab after much wary consideration. Not so with big trout. They dash for a lure like lightning, without careful observation, or any stopping to wonder what it is.

It is just because of these two opposite though characteristic habits that my floating nature lures will furnish anglers with new thrills, trebling the pleasure heretofore enjoyed.

With a stiff, regulation fly-rod, you can cast out any of these light nature fish-food imitations. You can play it at the surface; then in full view you can watch the gamy fish go for it and grab it. The lure cannot drop like a plummet to get snagged on the bottom; and even a tyro caster may place it among the weeds without trouble, till it is seen by the fish. There is no need of a rapid reel-in; and no bother of line tangle—which so often happens

ARTIFICIAL FROGS

when a heavy lure sinks to fasten its numerous double or treble hooks on a sunken tree-trunk. No weedless hook is needed for these lures.

It will be noticed that all nature lures are provided with but one single hook of good size. I consider treble hooks a bad feature, as only one barb takes hold on the fish.

With this preliminary, I will describe the frogs and the simple method required to use them. I consider the frogs my greatest achievement—so far —because of the difficulties to overcome in making movable legs and in making the shoulders appear above the surface yet having the body submerged.

In constructing the frogs I copied the green leopard frog for Eastern waters, the spotted brown pickerel frog for the Middle West, and the little red-bellied frog for the Pacific Coast States; though it is possible that one or the other may be found useful in different localities. The belly of the green frog is pure white, running to a bright yellow at the base. On the back, of bright green, are irregular spots of black. The pickerel frog has a white belly, spotted at the base with brown, and a brownish back spotted in black. The Pacific frog is white at the front, with bright red at the sides, running to scarlet spots at the base of the belly. All three weigh less than a quarter of an ounce.

The size and bulk of the frog are the only objections to its being used as a fly. Nevertheless, it can

NEW ARTIFICIAL NATURE LURES

be forced out, and cast by the lightest trout rod. While it rests on the water, the slightest agitation of the rod-tip will make the frog move its legs in the attitude of swimming. It is taken for granted that anglers know that with all artificial lures when grabbed, a rapid, though slight, wrist movement must at once be made to embed the hook. Natural bait is first captured, then held in the mouth and gorged at leisure. Only flies are gorged at once when taken.

I have a particular antipathy to that horrible method known as "trolling"; and I don't much enjoy or find sport in "still fishing"—when you sit in a boat on a lake, chuck overboard a lot of groundbait, then drop to the bottom a big night-walker worm, to shortly pull up a fat, lazy trout, which everybody says "can't be caught any other way," except on rare occasions in spring when they do sometimes rise to flies.

I do not believe these methods to be the "only way" to get fish from deep water. I know these frog and minnow lures will and have attracted fish to the surface of both lake and stream. Not always, perhaps, but most often. Yet in the event of the unusual happening, it is quite simple enough to place a buckshot four inches from the lure on the leader, to gradually sink near the bottom and play the lure in midwater.

To those unable to capture fish on lures or plugs

ARTIFICIAL FROGS

—the latter being an art very difficult to learn—or forced to buy live bait, perhaps catch frogs themselves, my lures will prove a real blessing. It is amazing what price some people will ask for live bait, especially at popular resorts and hotels. When you do get the frogs, after a single slap on the water the poor little beastie will refuse to move, swim, or kick; lying on its back it swells out like a balloon—and so would you and I, if sent swinging by our lips forty feet over the water.

Setting aside the undoubted cruelty of empaling live bait on the hook for a lure, how much easier it is to fish with a good lure that attracts the quarry—the only real, logical solution of the problem.

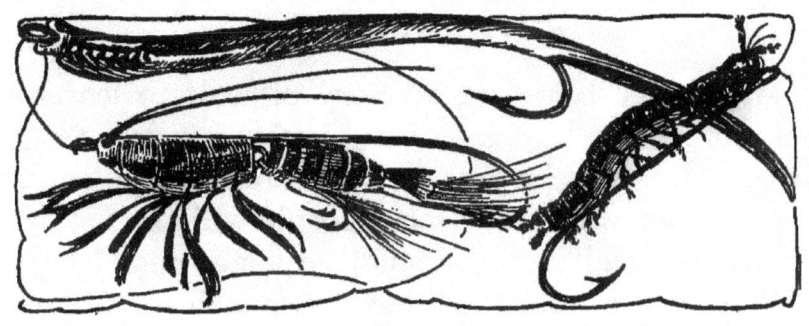

XVIII

THE THREE BEST NATURE LURES FOR BASS

The crawfish, helgramite, and lamper eel are exclusively bass food—pike, pickerel, and trout have little or no use for them. With a quantity of live baits on a fishing trip to lake or stream, I should be very confident of getting plenty of bass at any time, in any locality and under any weather condition.

Naturally they are most effective in river fishing, because they breed and live in rivers. On those three splendid bass rivers, the Schuylkill, Susquehanna, and Delaware, the crawfish and helgramite stand first in the heart of every angler who loves bass fishing; indeed, those three baits are universally used on those streams and on many other lesser bass streams. To my thinking, river fishing is in every respect infinitely more delightful than lake fishing, whatever method and baits are employed.

The helgramite is most popular for the reason

CRAWFISH: (a) *under view;* (b) *back view;* (c) *side view;* (d) HELGRAMITE, *side view*

THREE BEST NATURE LURES FOR BASS

that it is more easily captured, is very tough on the hook, lasts a long time alive in captivity if kept in a cool, dark place, and is always lively, anxious to get away from the hook.

The lamper eel is more delicate and soon dies. When dead it loses that fresh dark olive green to become a dull slate color; in that state bass do not take so kindly to it. It is hard to get; hard to keep fresh; and a perfect little devil in snagging your line. Digging lampers is far from a pleasant recreation on a hot afternoon. No wonder men and boys charge an average price of five cents each; and when you experience, as I have, the annoyance of seven out of ten being taken by chub (which usually abide along with bass) the price comes high to buy them. Yet, withal, bass love them very much indeed.

Crawfishing in brooks is altogether different; in fact, I consider it both amusing and interesting to catch crawfish. But it requires some experience to be a good crawfisher, for they are so nimble that they appear to crawl in every direction at the same time; so you have to be pretty spry and very patient to get a supply.

It would, however, be unwise on my part to describe how these baits are best caught, for I am hopeful that in the near future these nature lures will be found such effective substitutes as to be quite as good as, nay, better than, the natural bait,

NEW ARTIFICIAL NATURE LURES

all things considered. When that fortunate time arrives, game fish foods of all kinds will increase more rapidly, being left at peace to serve their purpose (as nature intended) of making game fishes more abundant and of greater size as time goes on. Whenever food is plentiful, game fish thrive and grow big.

I shall describe the nature crawfish first, for it is equally as good an imitation of the natural bait as the frog, if not an even better one. Also it is universally conceded to be the top-notch bass enticer.

Every intelligent bass fisherman knows that bass always swallow crawfish tail first, for the obvious reason that the claws fold together over the head as they slide down, and not spread out; otherwise the foolish bass would find Mr. Crawfish playfully pinching delicate parts of his gullet on the way down. For that very reason crawfish ought to be hooked (but never are) by inserting the hook point up through the belly, coming out at the back, half an inch below the eyes. Such a method immediately kills the crawfish, and that is the why most anglers, including myself, hook crawfish by the tail. In this manner, fish have much less chance to gorge the bait without getting entangled on the leader, especially if the hook is larger than the bait.

After many trials and a careful study as how best to place the hook to hold fast immediately bass strike at the crawfish's tail, I decided that the most

THREE BEST NATURE LURES FOR BASS

effective way would be to have the hook extend a little beyond the tail, so that a striking bass would swallow barb first: thus he would be made doubly sure of capture.

Glancing at the side view, anglers will notice the long, powerful hook, running from the eye on top of the body (entirely out of sight of the bass below), the curve ingeniously hidden by the crawfish's tail —made of turkey tail feathers and long hair of wolf. Underneath the tail is placed a double hook, set there to clinch the snap of the bass. This double hook is made invisible by hair from squirrel's tail. The body is a solid piece of painted cork, to which is securely fastened by silver wire the legs and claws, which are made of the long fibers of turkey tail feathers—the most pliable yet strongest material I can find for the purpose. At the base of the body the movable tail is fastened by a hinge. The tail is shown in the cut at its highest point, which is the position seen while in the water, though on a slack line the tail drops down, to move up again when a jerk is given to the lure. The horns are quills from a cock's hackle. They really are of no value to the lure, except to make it more lifelike to us. They should be nipped off by the fisherman to about half an inch.

It is only on very rare occasions that bass will rise from the bottom where they lie, to take a live crawfish at, or near, the surface. They invariably seize

NEW ARTIFICIAL NATURE LURES

a live crawfish while it is swimming near the bottom, going for it even after it begins to crawl along the bed of the river to hide under a stone. For that reason, sometimes, the same condition will prevail with the artificial lure, and to carry out the natural delusion it may be necessary to place one or two buckshot on the leader near the eye of the hook. If the water is sluggish, one shot is sufficient to keep the lure below, yet suspended above the bottom.

More success will be apparent if the bait is jerked and kept on the move. No full basket is gained by the sleepy angler, or one who stares around him without thinking what he is about. Bass are everlastingly cute; they are neither sleepy nor foolish.

Another way—especially good in swift water—is to fasten a dipsey sinker to the end of the line, then have the lure on a two-foot leader, which is attached to the line one foot above the sinker. The force of the water stops the lure from sinking to the bottom, but it floats at the same height as tied, according to the action of the water-flow. The sinker should be lifted now and then, to give life to the lure. This method applies also to the helgramite, the frog, and various floating minnows, to be tried only if bass fail to rise at the lure when near the surface.

The reason a live helgramite is so good to use as a bass lure is twofold. First, they are very tough; second, they are very active, swimming and wrig-

THREE BEST NATURE LURES FOR BASS

gling in the water all the time to get to their natural habitat, the bed of the river. Once there, they bore in the sand and so good-by. Therefore it is most necessary that the artificial be played and kept moving near the bottom all the time. I have made it of cork to float, hoping to induce bass to take it at the surface, or at least above midwater. Otherwise it must be made to sink with one or two small split shot fastened on the leader as before described.

The lamper eel has not been made to float because its peculiar shape and length of body would make a much more expensive lure if made of cork. It was found much easier to use a piece of rubber tube cut and bent to shape and carefully painted the life color. The long, powerful hook and the heavy rubber make it weigh somewhat more—a little over a quarter of an ounce; but it is quite light enough to cast with a long fly-rod. After casting, it should be allowed to sink near the bottom, then rapidly reeled in—its bent shape giving a peculiar twisting wriggle to imitate the motion of a live eel. It is quite possible (though not yet tried) that the addition of a small half-inch silver flat spoon attached to the eye of the hook would prove an extra attraction to the bass. Later on a trial will be made to construct a floating lamper of cork reenforced with silver wire and wound in silk. Such a lure will be more expensive; whether it will be worth while remains to be seen by a test.

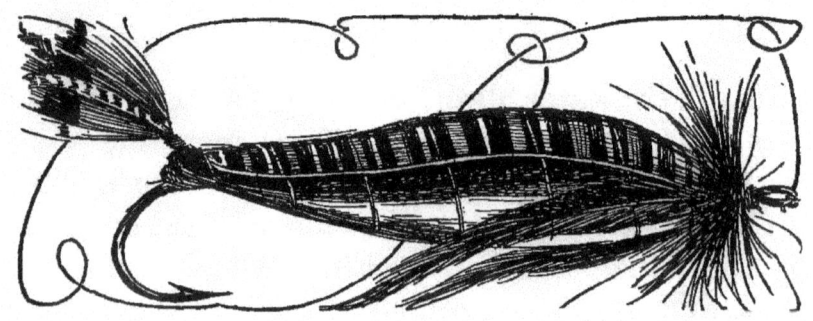

XIX

SILVER SHINER AND GOLDEN CHUB

NEW MINNOWS FOR SURFACE OR DEEP-WATER TROLLING AFTER BIG LAKE TROUT, TOGUE, MASCALONGE, OR SALMON

FROM different sections of this country—Maine, Nova Scotia, Ontario, Lake Keuka in northern New York—there have come requests for a nature lure in the shape of a shiner minnow to take the place of natural bait—a lure big enough for trolling in deep water after very large game fishes, running up to fifty pounds' weight. To meet such a demand, it was necessary to make numerous trials in order to overcome certain difficulties, the greatest of which was to get a bait equally good for use at the surface and at the bottom. I succeeded at last in producing what may be seen on this page— a silver dace or shiner—and a golden chub, constructed in various materials round a single power-

SILVER SHINER AND GOLDEN CHUB

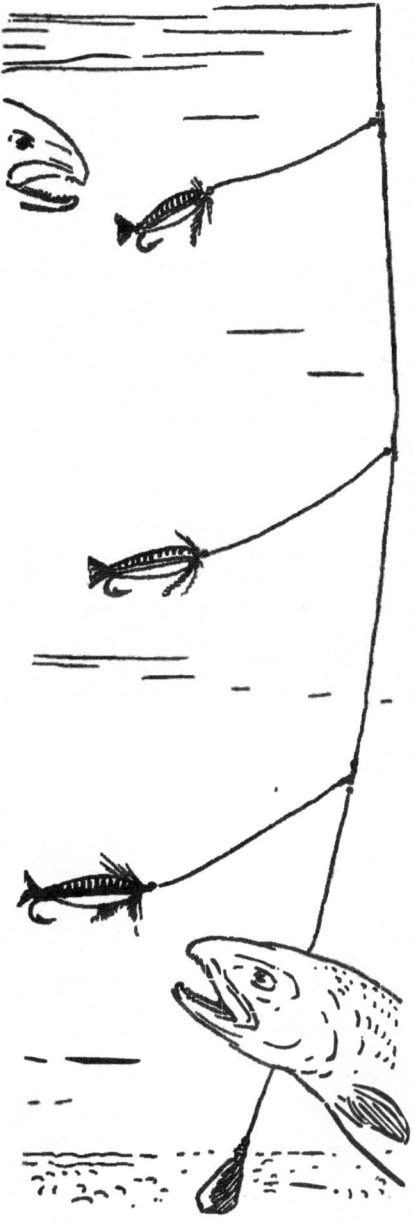

ful hook five inches long; making the lure measure from nose to tail tip six and a quarter inches, and more than one inch through from belly to back.

Three sizes are made—from nearly seven inches down to three inches. Each minnow is made in two parts: the back, of cork, is wound in dark blue and light green raffia, reinforced with silver tinsel; the belly, a solid strip of hard, heavy wood, cut to shape, is covered over with real leaf gold, or silver, which is varnished to retain its luster. Running along the middle body are three strands of bright blue, green and purple wool, well bound by strong silver wire. The side fins are

NEW ARTIFICIAL NATURE LURES

cock's hackles; and the tail is made of turkey's tail feather cut to shape.

From the illustration, anglers can judge only the form of this minnow; it is impossible to describe its beauty of color and truth to nature. The parts are so constructed and put together as to make the minnow swim upright, and glide through the water when trolled, exactly as if it were living bait; in addition, it has a buoyancy to float naturally whereever the sinker takes it, instead of dropping to lie lifeless at the bottom.

Every one of my baits heretofore has been made specially for "casting"—either at the surface or sunk—by a certain method to attain the best results. With this lure it has been found necessary to conform to the varied methods now in use in different localities for deep water trolling. A short description of these methods, accompanied by diagrams, may be useful to show how the minnow is attached to the line and the best way to capture these large game fishes.

VARIOUS METHODS OF TROLLING FOR LAKE TROUT

Spinning and trolling are carried on chiefly in large lakes where trout do not rise to the fly. The lake trout come to the surface very early in the spring, immediately after the ice melts; and the angler trolls for them on or near the top of the

SILVER SHINER (*actual size*) FOR EITHER SURFACE OR DEEP WATER TROLLING AFTER LAKE TROUT, MASCALONGE, PIKE AND SALMON

SILVER SHINER AND GOLDEN CHUB

water, the fish taking the lure viciously but rarely jumping into the air.

The proper tackle for surface trolling consists of a twelve-thread Ashaway cotton line, to which is attached a strong four-ply three-foot gut leader. On the leader you fasten one or two buckshot six inches apart. Use a good multiplying reel, and an eight-ounce rod, not longer than eight feet. If the trout run big, the large shiner is most seducing to them. It all depends upon the locality which bait is best, the silver or the gold; though I think they will strike viciously at both.

The deep-water troll requires more elaborate tackle than that used in surface fishing. Attach to the end of the reel line a cone-shaped sinker from three to sixteen ounces in weight, the size being dependent on the character of the bottom and the style of fishing preferred. If the bottom is jagged in shape, the line should be strong and the sinker comparatively small. The same holds good on smooth bottoms when fishing "slow and far off." If you prefer fishing with a short line, the sinker must be heavy. Few trollers use a rod: the line held in the hand enables you to be more sensitive to the slightest touch of the lead on the bottom, as your boatman rows slowly and regularly along. Nevertheless, a rod is much more valuable in playing the fish.

NEW ARTIFICIAL NATURE LURES

Three feet above the sinker attach a strong single or double-twisted leader (the average weight of the fish that are feeding should determine its strength) and two other leaders above the first, from six to ten feet apart, the distance to be judged by the depth at which the lake trout are taking the bait. Place swivels wherever needed; and let your sinker line be three feet long, and weaker than the reel line, so that in case of getting snagged among the bottom rocks you will lose only the sinker. Above all things, have the boatman row slowly along and with a cadenced movement. The secret of success lies in proper speed, the right depth, and the right place. As a rule, fish between late afternoon and dark.

TROLLING FOR MASCALONGE

From among the many methods, I have chosen the following as best suited to these new minnow lures. For some reason or other, the mascalonge is supposed not to be ready for live bait (minnows) until the fall. In the clear and swifter waters of the upper Ohio and its tributaries the mascalonge lies in the deep pools during summer and fall, where it is often taken by still fishing. But with these minnows it will be necessary to troll with a sinker light enough to be trolled slowly—similar to the methods practised on lakes—at various depths according to time and season, and where

SILVER SHINER AND GOLDEN CHUB

the fish are known to be. The best months are September and October; and the most favorable hours are early morning and late afternoon, though on dark and cloudy days, with a brisk wind, the middle of the day is just as favorable.

For short casting, row slowly along, in water from five to ten feet deep, and cast the minnow as near as possible to the edge of weed patches, reeling in again very slowly. When the wind and current are just right, it is a good plan to drift while casting. As soon as the fish strikes, and is well hooked, the boat should be moved to deeper, more open water by a skilled boatman; and care should be taken that the line is kept taut, in order to lessen the chances of the fish's taking to the weeds.

The minnow may be trolled along the edges of the channel, just outside the weed patches, from a moving boat, with a line of from thirty to fifty yards.

The tackle may be the same as that used for lake trout. Many anglers troll with hand lines of heavy, braided linen; but the use of a rod is of much greater service in playing large fish, should you succeed in getting them. A fish of such excellent game qualities deserves treatment of a better kind.

The range of these three splendid game fishes is so wide as to make it impossible to give even a short list of places. Lake trout, togue, and salmon trout may be caught in any of the thousands of big lakes through the Northwestern Hemisphere. The Pacific salmon does not take the fly; but magnificent fishing may be had in the salt water of Monterey, Santa Cruz, and Carmel Bays, where the method is to troll in thirty feet of water with smelt bait, of which this minnow is an excellent imitation. Fish of fifty pounds' weight are frequently caught; and their game qualities are equal to the salmon of Eastern waters. The mascalonge is well distributed throughout the Middle West and Canada, and is a worthy brother to the salmon.

If anglers will give these nature lures only half the effort they expend on live baits, they will enjoy sport enough to please the most fastidious. A small amount of good judgment as to where to get fish, how to get them, when to get them, and with what to get them, is certain to succeed. These lures will be found even better than live bait—certainly better than artificial lures heretofore tried.

There is no need for this big minnow to be made to revolve or spin. It glides along, or can be made to dart suddenly, just as the living shiner or smelt would do in its natural habitat.

To repeat what has been stated before: "Make a lifeless object a living thing: make the bait act alive by the ingenious manipulation of your line." If you are familiar with deep-water fishing, you

SILVER SHINER AND GOLDEN CHUB

will know of many better dodges used when live-bait fishing that will serve the same ends with these lures. In fact, *imagine* you are using a live bait; force the fish by your ingenuity to think the same; then it will go for it quickly enough.

I do not believe a spoon attachment of any size or make will add to this minnow's usefulness; though I know many anglers place spoons along with their live bait—which, by the way, is more often dead, and for that reason they have to make it spin. But this minnow swims along as if alive, and the brilliant sheen of the gold and silver bellies is sufficiently attractive.

Finally, in placing this giant minnow before brothers of the craft, I claim it to be a kindly, sportsmanlike lure, in place of what one of my correspondents terms "those murderous grappling irons offered to the multitude, which should be relegated to the use of municipal morgues."

No one living can feel more grief than I at the loss of a very large fish. One season I played a four-pound brown trout for half an hour. He was wilder than a captive wolf, with his leaps and lunges; and I fairly screamed with pain to see my leader snap like a bowstring on his last leap for freedom. Two days later I got him safe ashore, slowly and carefully working, till I was thoroughly exhausted when I slid on to the sandy beach. While making an examination of the contents of

NEW ARTIFICIAL NATURE LURES

this trout's stomach, I found nine hooks in various parts of its body, only two of which had been used with artificial flies.

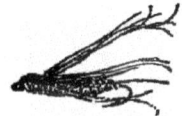

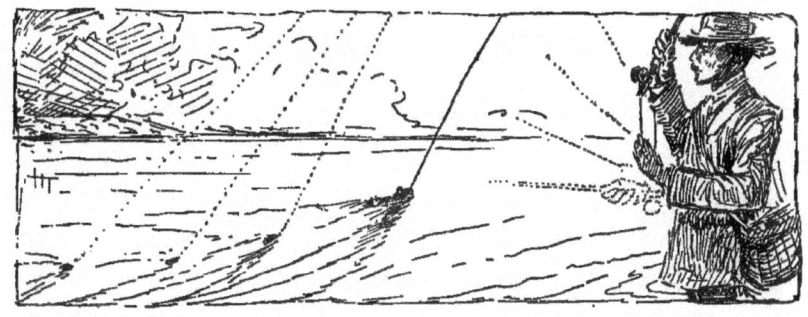

XX

THE RIGHT IMPLEMENTS AND METHODS

Their Importance for Sure Success

AMONG the numerous kindly letters received from anglers who have succeeded in catching fish with nature lures there are several from anglers who failed, and made inquiries as to why they did. There may be others having similar experience, and it is to these—if there be any—that I would like to point out a way to sure success. As all of my nature lures are made to float, they must of necessity be constructed of lightweight materials; and this makes it imperative that certain methods be strictly complied with for general success.

I am aware that most anglers act upon the assumption that they are skilful enough, when they get a lure, to know how to use it without instruction from anybody. But in the case of an entirely new idea—in fact, a complete revolution of existing methods—it seems to me advisable to take some notice of what the inventor has to say concerning the lures.

I must here call attention to the fact that the greater effectiveness of live bait (especially with bass) is in its continuous movement in the water. True, we see at times a minnow, frog, or crab lying perfectly motionless; but they dart off with considerable activity the moment they perceive a bass within their vision—a matter of self-preservation.

The same thing applies to imitation nature baits or lures—they must be made to act alive by the ingenious manipulation of the angler's rod-tip. This essential part of the method soon becomes mechanical, and the more you are expert in this practise, the sooner it takes the form of a greater delight in the game: viz., to make a lifeless object a living thing. No one can question that these lures are accurate imitations in form and color of live baits; and it rests with the angler to do his part in giving life movement to them. Every angler, if he tries hard enough, can make fish believe they take live bait.

It is astonishing how much can be done with a trout rod-tip in the manipulation of a lure or fly in imitating true to life the action of fish food. This is seen to perfection in dry-fishing—so far, the most perfectly artistic method in all fishing; and the very essence of it all is the perfection of the rod-

RIGHT IMPLEMENTS AND METHODS

tip play. This same thing I expect and truly hope anglers will attain with these nature lures.

Now, the first important thing is to use the right implements. You should have a powerful trout rod at least nine feet long (still longer is better), with a stiff yet pliable tip, at the end of which should be an agate guide and another one near the handle, a distance of nine inches from the handle; between these two the rest should be snake guides. You should have a good, yet soft, oiled-silk line that will slide through the guides as easily as if greased. I now use an imported tapered dry fly trout line, because I find it the best and I can, if desired, put on a fly without vexatious delay in changing reel and line. A two-dollar single-click reel will suffice.

The most important part of all is that you use with each and every lure a single bass gut leader from three to six feet long, neatly tied, without loop, through the eye of the hook at one end, the other end attached to the line with a loop.

One angler wrote that his frog persisted in floating on its back. Yet every frog is made the same, of material that so balances as to be impossible of itself to turn over when dropped into the water. I found out that the angler (an expert) had used a short double gut leader, only six inches long, which turned the frog every time he cast. Another used

NEW ARTIFICIAL NATURE LURES

no gut leader, but attached the light dragon-fly to a heavy, stiff oiled line which pulled the fly under the surface. Still another sat watching two hours with the rod resting on the boat, waiting for bass to grab the lure lying still at the surface. But he *did* get a strike when reeling in. These are not stupid, only careless, mistakes, due to not having read the instructions that go with each lure.

Every method, every bait, requires some particular kind of practise to attain success. The expert will get ten times more strikes with live bait than will the amateur who does not fish by method or rule. Casting the plug (Western style) is a case in point. I spent much time learning how to drop a plug properly, then to retrieve the line without snap or other troubles.

These nature lures are not made nor intended for trolling—except the larger sized minnows. But the frog, crawfish, grasshopper, helgramite, are all made suitable to cast the short distance of thirty feet, more or less.

If I have not in this book convinced anglers that nature lures (which exactly imitate the color, form, and life movements of natural baits) are equally advantageous to thoughtful anglers, to game fishes, and to the various creatures they feed on, it is not from lack of effort and years of patient study on my part. But I am confident that in time, with a

RIGHT IMPLEMENTS AND METHODS

little practise, nature lures will take their place in advance of, and succeed far better than, anything else, either natural or artificial. At least I myself shall make them do so, if others do not. It is only a matter of persistent effort in the right direction— that is to *make* the artificial *act* as the natural bait does. It seems to me such effort would furnish additional pleasure in the game of angling—to play artificial lures so skilfully as to deceive fish into believing them to be living things.

Moreover, the bass, trout, and pike angler, equipped with a complete set of nature lures, should be able to catch the bigger fish—at less trouble and expense—and to gain infinitely more sport, with keener delight, than he can obtain with any other lures, natural or artificial.

THE END

www.ingramcontent.com/pod-product-compliance
Lightning Source LLC
Chambersburg PA
CBHW061256110426
42742CB00012BA/1931